THE WORST YEAR EVER: 536 AD

Volcano, Mini Ice Age, Plague, and an Emperor Who Taxed the Dead

REECE KIMBLE

Printed in the United States of America
First Printing 2022
First Edition 2022

10 9 8 7 6 5 4 3 2 1

To our ancestors who survived
so that we may enjoy life.
&
To our beautiful Earth -
may it be our forever home.

TABLE OF CONTENTS

Introduction

Why Was 536 So Bad?

In 536 AD, a huge volcanic eruption changed the course of human life as they knew it. One volcano impacted nearly every part of the globe. It was so big that scientists only recently figured out exactly which volcano erupted and where. The global impact was devastating—and that was just the beginning.

Following the volcano, an impenetrable darkness covered most of Europe and Asia for over a year. Historian Procopius wrote that the sun gave out "light without brightness, like the moon" for the whole year (Gibbons, 2018).

The lack of sunlight led to a period of intense global cooling which we now call the Mini Ice Age. Crops failed globally, leading to widespread famine and starvation.

But our ancestors were under even more stress than that. Imagine what it would be like to be plunged into a sudden darkness, dealing with winter conditions in the middle of summer. Even with modern science, this would be incredibly stressful and confusing. For contemporary people, without the same understanding of science, it was often seen as a sign of the end of the world.

People were desperate, and the combination of stress, unrest, and famine led to a period of intense political change and turmoil. It was the beginning of the end for the Eastern Roman Empire and a time of intense conflict in Europe and Asia.

To make matters worse, the near-global famine left humanity vulnerable to disease. The plague appeared shortly after, killing approximately 5,000 people every day in Constantinople alone (Mordechai & Eisenberg, 2019).

The destruction and loss that happened in 536 AD and following was shocking and catastrophic. When we hear about these events, it's easy to put them into perspective with other disasters. However, it's important to understand the gravity of these events to the people of 536 AD.

For them, this time period would have felt unprecedented, scary, and overwhelming… not unlike what we've been experiencing since 2020. In 536 AD, one bad thing led to a downward spiral that took almost a century for most of Europe to recover from.

People around the globe were dealing with the loss of their family and friends to the plague on top of famine, war, and brutal weather conditions. It would have been incredibly stressful to live through a period of such intense loss. We need to be empathetic to understand contemporaries' circumstances and how they connect with our own.

So what happened in 536 AD? How did one volcano lead to global cooling, famine, the plague, and political unrest?

In this book, you'll discover the exact chain of events that occurred in 536 AD, why they happened, and how they were connected. You'll learn how these events connect with modern crises like the COVID-19 pandemic and climate change.

What can we learn from their mistakes, and how can we change the course of our future?

To begin to unpack this question, we have to understand what happened in 536 AD, beginning with the volcanic eruption that started it all.

Chapter 1

The Volcanic Eruption

The volcanic eruption that occurred in 536 AD had a widespread impact on the global climate, food, health, and even politics. It truly changed the world.

However, we didn't know that an eruption caused all this until only a few decades ago. Contemporaries in 536 AD wrote about a sudden cold in the middle of summer and a strange fog that blocked out the sun, lasting for months, but we had no proof of these events. Plus, we had no certain way of explaining them, so nothing came of it until recently.

In the 1990s, scientists finally found proof of these unusually cold summers using dendrochronology, the study of tree rings. Using tree rings, specialists can see how much a tree grew during a certain period of time. Sure enough, tree ring analysis proved that there was little to no tree growth during the sixth century due to a lack of sunlight or rainfall (Helama et al., 2018). But what could have caused this lack of sunlight or rainfall over such a long time span?

It would still take years before scientists finally uncovered the cause of these cold, dark summers using a different technique: ice core analysis.

New, precise ice core analysis techniques can detect chemicals and compounds in the ice down to the month—even 2,000 years in the past (Gibbons, 2018). Certain compounds are always released by volcanoes during an eruption, and these were present in the analysis of ice cores. Things like sulfate, volcanic glass, and sulfuric acid deposits proved that there was an eruption.

Where exactly was this eruption, and how do we know for sure that's what happened?

Finding the Culprit

Since this eruption had such a profound global impact, that question is harder to answer than you might think. Initially, scientists had a huge range of theories about the suspect, ranging from a volcano in El Salvador to one in Indonesia and even dismissing the volcano theory entirely.

Krakatoa, Comets, and More

Some theorists believed that the Krakatoa volcano in Indonesia caused the worst year ever. The volcano was known for its repeated, intense eruptions, making it an easy target. Plus, for a short time, historians believed that it erupted around the same time as the global cooling occurred. Unfortunately, these claims were quickly debunked as there was no concrete evidence that the volcano erupted in 536 AD.

Similarly, researchers in the 1980s thought that the Rabaul volcano in Papua New Guinea might have been the cause. This was mainly because of the volcano's tropical location, which would have easily spread ashes over Europe and Asia. Dating techniques at the time suggested that the volcano erupted around 536, but this was disproved later. We now know that the volcano erupted in the late 600s or early 700s (Newfield, 2016).

Later on, another theory claimed that multiple comets might explain the effects. Interestingly, even a relatively small comet fragment (just over a quarter mile) could cause the "dust veil" that we saw in 536, alongside a period of global cooling (Rigby et al., 2004). All of this data was theoretical, and there was no proof that a comet struck around this time period. Initially, that didn't matter. Scientists reasoned that the comet might have struck in the middle of the ocean or been burned up in an airburst, so we wouldn't have found the crater. Small comet collisions are quite common, so this theory gained popularity for a while in the early 2000s. It was eventually disproved when ice core evidence of the 536 AD volcano was found.

The Ilopango Theory

The Ilopango volcano in El Salvador was believed to be one of the biggest volcanic eruptions of all time, and scientists thought it happened just before this period of global cooling. This made it the perfect trigger for the events of 536 AD.

This theory was strengthened when new evidence came up in 2008, suggesting that a volcano near the equator caused the 536 AD eruption (Larsen et al., 2008). Scientists believed that the sulfate deposits found in Greenland and Antarctica, alongside tree ring analysis, proved that the volcano was big enough to cause the effects we saw in 536 AD *and* would have been in a tropical location.

Interestingly, later research dated the eruption in 539 or 540—not in 536 AD as initially thought (Greshko, 2019). This was later disproved, but another volcano did erupt around 540 AD (more on this later). This is important for us because the second eruption prolonged the initial effects, including global cooling. The effects of the first volcano wouldn't have persisted, or had such a drastic effect, without the second eruption.

More recent research dates the Ilopango eruption to around 431 AD (Smith et al., 2020). So, while the Ilopango volcano erupted at a similar time and had a huge global effect, it wasn't the eruption of 536 AD.

Modern Research

Using advances in ice core and tree ring analysis, scientists have narrowed down where the volcano was. It actually wasn't a tropical volcano at all! Recent ice core analysis proves that it was a massive eruption somewhere in Iceland (Loveluck et al., 2018). We still aren't sure exactly which volcano it was, since the eruption had a devastating toll on the entire area, but hopefully future research can uncover this.

You might be surprised by how difficult it was for scientists to figure out what happened in 536. Considering the massive global impact, you might think that it would be easy to determine where it started, but it's actually the opposite. It's harder to tell where something occurred when it was big enough to impact the whole world. It'll have a colossal impact on its region of origin overall. Pinpointing an exact location is getting easier using modern scientific advancements, but there are still plenty of roadblocks in figuring out exactly which volcano erupted.

We do know that the eruption in Iceland would have had a devastating impact in the immediate area, alongside global consequences. The eruption sent ash and volcanic dust high into the atmosphere, creating the strange fog that contemporaries wrote about.

So how did this volcanic eruption lead to all the negative effects we touched on in the introduction?

The Volcanic Winter

When you think about a volcanic eruption, you probably imagine spewing lava and clouds of ash. People living there would be impacted by the event, but it wouldn't have a global effect… right?

Volcanoes can do more damage than you might think. An eruption can cause global cooling, which can lead to a chain of bad events including crop failure, starvation, and sickness.

The effects of a volcanic eruption can be similar to what we would see after a nuclear war.

If you've ever heard of the term *nuclear winter*, you'll know what I mean.

A nuclear winter is what we would expect to happen if there was a large-scale nuclear war. The soot and dust from the explosion and from resulting fires and firestorms would be blasted into upper areas of the atmosphere, creating a layer that blocks out the sun. The lack of sunlight leads to global cooling, creating winter conditions even in the middle of summer (Robock & Toon, 2016).

This is almost exactly how a volcanic winter works. When a massive volcanic eruption happens, it produces large amounts of ash and airborne sulfur which are released into the upper layers of the atmosphere. These compounds reflect sunlight, meaning that less sunlight can get through to Earth. It's all reflected back by the sulfur and other compounds. The higher up these compounds get, the less likely they are to be washed away by rainfall. These materials can take years to disperse on their own.

This is what scientists believe happened in 536 AD: a massive volcanic eruption caused a volcanic winter in the middle of summer, which led to constant cold, low precipitation, and prolonged cloud cover and fog.

Droughts

As if the sudden global cooling wasn't bad enough, volcanic eruptions also tend to cause a lack of rainfall.

Basically, the airborne volcanic compounds that block out the sun also keep the water cycle from happening normally. Blocking out the sun reduces the amount of radiation that can get through to the ground. Less heat means less water will evaporate, meaning that there is less water in the air to condense into clouds and cause rainfall. This effect gets even worse when you factor in the sudden cooling.

In 536 AD, droughts were recorded across Europe and even in Peru (Keys, 2000).

Droughts combined with the cooling effect to ruin crops and create a global famine.

Flooding

Meanwhile, in China, snowfall in the mid-August ruined the harvest (Keys, 2000). So why was there *snow* in China when other parts of the world were struggling through a drought?

It has to do with the way the atmosphere cools after an eruption. When there's a volcanic eruption, the worldwide cooling doesn't happen evenly. It actually gets cooler over land masses than it does over oceans, which creates massive flooding in some regions and uncharacteristic droughts in others.

This made matters worse, as even regions that weren't in a drought couldn't reap their harvest. They were still at the mercy of dangerous, unpredictable weather.

Famine

Extreme cold, droughts, flooding, and lack of sunlight made growing food incredibly difficult. Most harvests failed. In fact, a "failure of bread" was reported in parts of Europe from 536–539 AD (Gibbons, 2018). This meant that most people were forced to live off of any food they had saved from previous harvests, which often wasn't very much—if anything at all.

Whole societies struggled to feed their people, causing mass starvation. Those who survived were left weak and vulnerable to disease while all in a period of constant cold.

That's partially why the plague took such a toll on Europe and Asia.

It's very common in history to see a plague shortly after a volcanic eruption because of the bad chain of events it causes. When crops fail and people starve, it's easier for them to get sick and get the people around them sick, too.

Lack of modern medicine didn't help, as people didn't know the best ways to protect themselves from getting infected. The poor nutrition intensified all of these factors, leaving people even more vulnerable.

The Emotional Toll

One historian described the event as, "A winter without storms, a spring without mildness, and a summer without heat" (Cassiodorus, 1886). He even said that the sun seemed to be in eclipse all year.

Can you imagine what it would be like to live through something like this? Naturally, it had profound impacts on the well-being of contemporaries. The sudden cooling and drought caused crops to fail, which were a crucial source of food. Since the cooling effects were seen almost globally, they couldn't be handled using existing trade routes.

The volcanic eruption caused a perfect storm which made survival incredibly difficult for people worldwide, and that's only the practical implications. Imagine how scary it would be to wonder if you or your family would survive the winter, not knowing when the winter would even end.

The stress of these experiences was only amplified by the lack of scientific knowledge. Remember that people at the time would have had no reasonable explanation for why all of these things were happening. Due to the strong religious influence and lack of information, we can see why contemporaries would have thought the world was ending.

This situation has some alarming parallels with modern times. Luckily, we have much more advanced science and technology to understand our circumstances. However, we've also dealt with increasing food shortages and a global pandemic, as well as a climate crisis. It's been proven that global warming leads to more frequent and intense volcanic eruptions, alongside other natural disasters. If we continue to go down this path, things can only get worse. In 536 AD, it shows us what our future could look like. That's why it's so important to understand what happened and why. We want to keep history from repeating itself.

As you're about to see, things only got worse from here. If you thought 2020 was bad (and it was!), you might be surprised to hear how much worse it can get. Read on to feel a little bit better about modern times. Even when we're surviving horrible, scary circumstances, it could always be worse than it is.

For example, think about the volcanic eruption that happened January 15th, 2022 in Tonga. In the satellite footage of the eruption, it looks *huge* even though satellites literally captured it from space. The explosion was big enough to cause a tsunami and completely ravage the surrounding area, covering whole towns in ash. This was a horrible disaster for the people of Tonga, and approximately 85% of the population was directly impacted by the event (United Nations, 2022). A lot of these people were rendered homeless and struggled to access even the basics, like clean drinking water. As a nation, Tonga will be recovering for years.

However, even this disaster wasn't big enough to cause the global impact we saw in 536 AD. While Tonga was propelled into an immediate state of disaster and is still recovering, the rest of the world has gone back to normal relatively quickly. We didn't experience any huge temperature shifts or a constant fog. This will hopefully give you some perspective on just how massive the Icelandic eruption was in 536 AD.

CHAPTER 2

MINI ICE AGE

The volcanic eruption was big enough to send the world into a sudden volcanic winter… but it was an even bigger deal than that. The eruption marked the start of the coldest decade in 2000 years (Gibbons, 2018). That's because the eruption triggered global cooling that didn't just go away after a year or when the fog eventually cleared. It lasted for over 20 years—from 536 AD to about 560 AD (Peregrine, 2020).

These drastic temperature changes made life even more difficult for contemporaries. People struggled to grow food and survive the harsh conditions. Thousands of people lost their lives, and those who survived dealt with increasing unrest, the plague, and political conflicts.

THE CAUSE OF THE MINI ICE AGE

Was this volcanic eruption truly the cause of the Mini Ice Age, or was something else to blame?

It turns out that the eruption changed the world in more ways than we initially thought. Modern evidence shows that the Icelandic eruption was responsible for the start of the Mini Ice Age.

The "Dust Veil" and Sea Ice

You learned in the previous chapter that the volcanic dust and ash propelled into the atmosphere during an eruption have a direct effect on the water cycle. Less heat reaches the Earth's surface, meaning that less water evaporates to form clouds and cause rainfall. Well, this also means that more ice builds up on the sea. As sea ice continues to grow, it makes it more and more likely that the area will stay cold.

It's the same thing when you have a ton of snow in your backyard. It takes longer exposure to warmer conditions for two feet of snow to melt than for two inches. Basically, the more that the ice built up, the harder it was for it to melt. The ice began to sustain itself, which, in turn, sustained the cooler temperatures.

In fact, researchers theorize that the decrease in solar radiation wasn't even necessary to cause the global cooling effect: The sea ice alone was enough to have drastic impact (Miller et al., 2012).

Unfortunately, this wasn't the case in 536 AD: It wasn't *just* sea ice at work. The increased sea ice was combined with decreased radiation, clouds of ash, and airborne sulfur that all contributed to global cooling in 536.

However, the Mini Ice Age likely wouldn't have happened without the eruptions that followed in 540 and 547. The huge, repeated volcanic eruptions enhanced and prolonged the global cooling effect, taking it from a volcanic winter to the Mini Ice Age.

We know from the last chapter that large volcanic eruptions and global cooling go hand in hand. In fact, this is almost the same scenario that happened much later after the Tambora eruption in 1815.

The Tambora Eruption: Volcanoes and Mini Ice Ages

The 1815 Tambora eruption is known as the most massive and destructive volcanic eruption in history (Encyclopedia Britannica, 2018). Mount Tambora is located in Indonesia, and people on surrounding islands were hit the hardest. Ash in the stratosphere blocked out the sun, and combined with global cooling, it became impossible to grow crops in the region. Thousands of people died due to famine and sickness.

Even Europe and North America experienced the aftermath of the eruption, including failed harvests and famines. Frost and snow in the middle of summer were partially to blame, leading the year to be infamously known as "the year without summer" (Encyclopedia Britannica, 2018).

Unfortunately, the bad news doesn't end there. All of this happened at the tail end of yet another Little Ice Age, this one occurring from about 1300 to 1850. This eruption and other large eruptions at the time prolonged the Little Ice Age, making it last even longer.

This is eerily similar to what occurred in 536 AD. We can only imagine the impact that the eruption would have had on the surrounding area, and the fact that it spread worldwide is proof of that.

What's most interesting is that volcanoes have been repeatedly connected with mini ice ages. The 536 AD event was not a unique occurrence but proof of a larger pattern. For the people living at the time, though, it would have felt completely foreign and horrifying.

A Period of Prolonged Suffering

The ongoing natural disasters made 536 AD and the years after an incredibly difficult time to be alive. In almost every region, people were fighting to feed themselves and their families, to survive the cold, and just to make it to the next day.

Let's get into what was going on for contemporaries during the Mini Ice Age.

Famine, War, and Unrest

The Mini Ice Age caused crops to fail all over the globe. Early frost, sudden temperature changes, and heavy snowfall were huge contributors. The unrelenting cold meant there was no relief in sight.

Times of suffering commonly lead to war and civil unrest because they exacerbate existing tensions. This can happen on an individual level and a political level.

For example, say you don't get along well with your next-door neighbor. In average times, you can keep the tension from escalating into anything serious. But, when you're both starving and they steal your last loaf of bread, you might not be feeling so generous.

That's exactly what happens on a broader scale, too. If you don't get along with a neighboring nation, that tension will only be increased by famine, especially if you have nothing to spare for them in their time of need. Stress is already high, and people are more likely to say and do things they'll regret. This can lead to a full-blown conflict.

Some political leaders saw the famine as a way to gain an advantage. If their people were weak and starving, it meant that their enemy was weak and starving, too. Some leaders attacked when their own people were vulnerable because they were desperate to claim more territory or obtain some other goal.

On another note, people are more likely to get frustrated with their leadership when they're going through bad times.

That's because times of strife amplify poor leadership. It can feel like the leader has less impact on the community when times are good because they don't necessarily need to do much to keep the population happy. However, when people are freezing and starving, it will become obvious if the leader can't (or won't) help their people.

Uncertainty and economic instability can also create civil unrest. When people are poor, unsafe, and unsatisfied, their leaders' luxurious lifestyles will come under greater scrutiny. Plus, leaders might impose higher taxes or other regulations to deal with the economic issues, frustrating the people even more.

Times of strife highlight issues that were already present and bring them to the forefront. They also create new problems that require urgent action. When leaders fail to support their people, unrest is natural.

Cold and the Plague

The Mini Ice Age was a huge contributor to the plague. As you can imagine, the sudden drop in temperatures was difficult for the people as well as the crops. Together, the famine and cold left people vulnerable to sickness.

That wasn't the only way that global cooling worsened the plague, though. The Mini Ice Age also caused changes in rat movements across Europe and Asia as they tried to avoid the cold (Sarchet, 2016). Since the plague-causing bacteria was carried by rats and fleas, this worsened the pandemic.

The plague was first brought to Constantinople by boats arriving from Egypt. Contemporaries in Constantinople relied on imports, especially imported grain from Egypt, to sustain themselves. The rats thrived in the Egyptian granaries and were often carried over in grain shipments. Global cooling could have led to more or different rats in the grain shipment, some of which were carrying the bacteria that caused the plague (Mark, 2020).

Since it was so cold and people were starving, they ended up spending a lot of time indoors. The rats sought out warmth and often ended up in people's heated homes, so people were even more likely to get exposed to the bacteria.

Plus, we know that the cold killed crops and ruined harvests all across Europe, the Americas, and Asia. This meant that any remaining food was likely to be eaten—even if it was contaminated or poor quality. This exposed even more people to the plague.

Declines in Art and Culture: The Dark Ages

In pop culture, this time is frequently referred to as the Dark Ages. This title might seem fitting, considering the clouds of ash that blocked out the sun—but it's an inaccurate term that scientists today shy away from. Viewing this time as the Dark Ages implies that something was "backward" or wrong with people during this time, especially compared to periods like the Enlightenment and Romanticism. Early scholars used this term to refer to a time without knowledge, where people were "in the dark."

However, this assumption is a disservice to the people who survived these trying times. We have no reason to believe that sixth-century people weren't creating and enjoying art and literature. Until recently, we just didn't have proof of their creations (Hughes, 2021). Now, we have collections of art, pottery, and other cultural artifacts that prove the creativity of contemporary people. These artifacts reflect rich cultures, stories, and histories, including references to constellations, gods, and goddesses (Metropolitan Museum of Art, 2022). Many of the things they made were lost to time, but that doesn't mean they were never created.

Culture, art, and literature actually thrived during the Middle Ages, although not necessarily during 536 AD (Hughes, 2021). With the sudden temperature drop, near-global famine, and the plague following shortly after, people hardly had time for art and culture. There may have been a brief "artistic recession" as contemporaries focused on feeding themselves, surviving the cold, and staying healthy amid an outbreak.

Art and culture tend to falter during difficult times in human history. People care less about having beautiful buildings and exciting events when they're struggling to find food and clean water. That doesn't mean that the people during this time were "backwards"; it's just that they were going through an incredibly difficult time. We should give contemporaries a break. Would you create incredible art and literature while starving during a sudden winter in the middle of summer? I certainly wouldn't!

The sixth to tenth centuries have also been referred to as the Dark Ages due to the reigning influence of religion rather than science. More modern societies shunned the people of the Early Middle Ages for their religious focus. This is an unfair judgment on contemporary people who didn't have modern science to rely on. It's important to recognize that religion would have informed most people's beliefs and values at the time without judging them for that. They did the best they could with the circumstances they had.

Alternatively, some historians appreciate the term Dark Ages because it highlights how little we know about this time period. Our records are few and far between, so it's hard to make decisive conclusions about historical events. In that respect, the Dark Ages is a fitting term. However, it still promotes an unempathetic view of contemporary people that focuses on what they lacked rather than what they knew and went through.

For that reason, even though 536 AD was a devastating time that we don't know much else about, I don't call it the Dark Ages.

The Perfect Storm

Combining these factors together, we can see why 536 AD was such an awful year. Like dominos, one disaster caused a chain reaction until every aspect of life was affected. People were cold, starving, and sick. It seemed like the constant winter would never end, which only heightened anxieties and political unrest.

All of these factors led to declines in art and culture as people struggled to survive. This was the reality for thousands upon thousands of people in the sixth century. Even those who weren't directly impacted by the eruption felt its aftereffects, and they were devastating.

However, this actually wasn't the case for everyone. Some regions flourished due to the eruption and resulting climate changes.

The Arab Peninsula

The Arab Peninsula, now Saudi Arabia and its surrounding countries, was one of the few regions in the world that benefited from global cooling. Just as the eruption created a string of bad events in most of Europe and Asia, it created a positive domino effect in the Arab Peninsula.

Increased Rainfall

The Arab Peninsula saw increases in rainfall due to the climate changes. While this led to flooding in a lot of other places, this region got lucky. They didn't seem to experience much (if any) flooding. Instead, the cooling and reduced evaporation meant that the typically dry region was able to sustain more plant life (Büntgen et al., 2016).

This had some obvious benefits like improved harvests, but it improved other aspects of life, too. The cooler temperatures and increase in plant life meant that travel was easier, and people could travel farther. Camels were often used in these regions for transportation, and they could feed on the vegetation to travel longer distances (Büntgen et al., 2016).

The increased rainfall also meant that more regions were inhabitable, so people could use more of the land. Empires could expand, people could spread out, and more buildings could be made to support them.

Rise of the Islamic Empire

All of these benefits meant that leaders in the Arab peninsula had more resources at their disposal. Growing their empire got easier as the rest of the world experienced a huge setback.

This gave them a window of opportunity. Nearly all empires were striving to grow their borders and conquer other nations, and the Islamic Empire was no different. Greater plant life and cooler temperatures sustained the movement of Arab armies and their supplies, allowing them to travel more easily into other nations and attack them while they were weakened. This was how Islam became an established religion and is partially why their conquests were so successful (Büntgen et al., 2016).

Downsides

Unfortunately, it wasn't all sunshine and rainbows for the Arab Peninsula. Contemporaries in this region still experienced setbacks, and although they weren't as dramatic as they were in some other regions, they were still debilitating for the growth of the nation and devastating for the people.

While they may have avoided famine and drought, they couldn't avoid the plague. International trade was an important aspect of life, even in the 500s, and diseases spread quickly along trade routes.

A crucial dam also collapsed in the Arab Peninsula during the sixth century, which could have been an aftereffect of the eruption (Paowary, 2018). The Marib Dam was an important structure for the people of the Arab

Peninsula, which allowed them to grow crops in what would normally be a desert. The city of Marib flourished nearby. Marib was an economic hub and home to the Sabaean kingdom, making it an important destination for merchants and travelers.

Unfortunately, the structure fell into disrepair and eventually collapsed in 570 AD. The collapse of the dam made it difficult to grow crops in the area and forced people to evacuate the city.

Historians are still unsure about the exact cause of the dam's collapse, but it might have been due to an earthquake or heavy rain (Paowary, 2018). Either of these could have been caused by the Icelandic eruption.

Studies have shown that volcanoes and earthquakes tend to cause each other (Australian Museum, 2021). A huge volcanic eruption can trigger a shift in tectonic plates, which could eventually cause an earthquake even as far as the Arab Peninsula. Similarly, heavy rains could have been brought on by the pressure and temperature changes as a result of the eruptions. Whether or not this particular event is related to the eruption, the possibility is intriguing.

The collapse of the Marib Dam and ongoing impact of the plague made times difficult for people in the Arab Peninsula. They may have been flourishing compared to Europe and Asia, but they still had their share of difficulties.

The Mini Ice Age: Drastic Climate and Cultural Change

The sixth-century Mini Ice Age paved the way for dramatic changes across the globe. While the sudden climate shifts wreaked havoc on some parts of the world, others were finally given a chance to flourish. This started a period of political and cultural change as nations worldwide grappled with their new circumstances.

For some leaders, this would mean fighting to sustain their people and their power. For others, it was a welcome opportunity to grow and develop.

In the next chapter, we'll dive into how these changes played out for the Eastern Roman Empire. How did they succeed, and why did they eventually fall? All the details, including the "golden age" and the emperor who taxed the dead, are to come.

CHAPTER 3

FALL OF THE ROMAN EMPIRE

THE WESTERN ROMAN EMPIRE

By 536 AD, the fate of the Western Roman Empire had already been sealed. It had been steadily declining for years due to ongoing conflicts with neighboring groups, like the Germanic tribes and the Goths, before finally falling in 476 AD (Andrews, 2019). These conflicts weakened the empire until it finally collapsed after multiple raids and attacks. This took a huge toll on the empire's population, who were exhausted from constant war.

It didn't help that the Western Roman Empire had gotten so big. The sheer extent of land made it difficult to sustain connection between the regions, which became a big problem during invasions or disputes.

Pop culture tends to see Rome through rose-colored glasses, but it wasn't an ideal place. This was especially true as it neared its downfall. The city relied heavily on slave labor from conquered nations. As the empire failed to bring home new manpower, this took a toll on its inner workings. It couldn't produce as much, leading to economic issues alongside the failed military campaigns (Andrews, 2019).

Before its collapse, the empire was in constant civil war. Leaders were pressured to invest in the military to maintain order and fight off constant invasions, meaning that less could be spent on maintenance and infrastructure. This only worsened conditions inside the city and led to more civil unrest.

When the Roman Empire was divided into two sides, the East and West, this worsened the problem. This was supposed to make it easier to govern the population but eventually made the two sides drift apart. While the prosperous Eastern Empire grew, the Western side struggled. By this point, the two sides had very little in common and even spoke different languages (Andrews, 2019). While the Eastern Empire was well protected, the Western side wasn't. Attacks that failed on the Eastern side would naturally move to the West where they would be more successful. The Eastern Roman Empire simply couldn't do enough to sustain the Western side despite multiple attempts.

THE EASTERN ROMAN EMPIRE: PRE-536 AD

While the Western Roman Empire crumbled, its Eastern counterpart thrived—at least, for a while. Their society was open to foreigners, even the people they considered "barbarians," as long as they would get baptized and pledge loyalty to the emperor (Teall

& Nicol, 2018). So, even though the empire is known for being homogenous, it was mostly homogenous in language and culture—not in the ethnic background of its people. In its earlier years, it was a surprisingly mobile social system, but this changed as the years went on.

Unfortunately, the events of 536 AD marked the beginning of the end for the Eastern Roman Empire, as well.

The Decline of the Eastern Roman Empire

The aftereffects of the eruption took toll on all of Europe, but the Eastern Roman Empire (or Byzantine Empire) was hit particularly hard. It's an effective case study in the eruption's complex web of consequences.

Famine and Cold

The resulting crop failures had a dramatic impact on the empire, particularly the "failure of bread" from 536–539 (Gibbons, 2018). Crop failure across Europe meant more reliance on imports, and many importing countries were also struggling with the harvest. When the second volcanic eruption occurred in 540 AD, it extended the cold snap and prolonged poor agricultural conditions.

Byzantine historian Procopius neatly sums up how the political conflicts connected with the natural disasters. He says, of the darkness caused by the eruption, that it seemed like "the sun in eclipse," and that, from that time, "men were free neither from war nor pestilence nor any other thing leading to death" (Procopius, 1916). His perspective shows that even at the time, people connected the aftereffects of the eruption (although they wouldn't have known what they were) with war, conflict, and death.

This was partially due to the divide, particularly in religious texts, between light and dark (Tunturi, 2011). Even today, we see the same themes in our media. Regardless, it would be natural to feel uneasy when the sun suddenly seems dimmer, and you're under a constant fog. Contemporaries saw the fading sun as a bad omen and a sign of the end of the world. While they didn't quite understand how the dim sunlight connected with their troubles, their assumption of a bad omen was pretty accurate.

Unfortunately, the Eastern Roman Empire was no exception when it came to the connection between famine, tension, and unrest.

At the time of the eruption, the Eastern Empire was fighting the Gothic War. After the fall of the Western Roman Empire, they were trying to reclaim the lost territory. However, the sudden darkening of the sun and cold snap made things difficult for both sides. The troops were prepared for a summer mission—not a winter one. When snow fell in the middle of summer, the armies were both physically unprepared and mentally unready. The unending winter made the mission much more difficult than they had expected.

The Plague: Effects on the People and the War

The Byzantine Empire was hit hard by the plague outbreak, losing between one-third to one-half of its population to the disease (Gibbons, 2018). This was devastating for the empire as a whole and for the individual. Imagine living through an outbreak that kills between one-third to one-half of the people you know. Those who survived the plague lost friends, neighbors, and loved ones.

At its worst, there was no room or time to provide proper burials, meaning that infected bodies were left in the streets. This was both emotionally devastating and unsafe. Survivors were unable to give their loved ones proper goodbyes, *and* they were at an increased risk of infection themselves.

The survivors emerged from the outbreak exhausted and weak from taking care of the sick. They were vulnerable to illness and starvation, especially if they had no surviving relatives or loved ones to look after them.

There were less able-bodied people to clean up the streets, bury the dead, process imported goods, and manage crops. This made bad times even worse for the Byzantine Empire. Survivors struggled to maintain cities and towns ravaged by the plague.

Plus, population loss meant less troops to send to the front lines of the Gothic War. This meant that what might've been a quick victory was drawn out, depleting the empire's resources and drawing their attention away from other matters. The Eastern Roman Empire weakened as they were forced to concede territory in other regions to continue fighting to reconquer Rome (Rosen, 2007).

For example, the Roman-Persian Wars were greatly influenced by the plague and the Gothic War, which left the empire weak to invasion by the Persians. They successfully invaded and destroyed several cities. This was both costly to repair and devastating to the population.

The onset of the plague also marked a major turning point for the army in Italy. The troops were set to reconquer Rome, which had been occupied by the Ostrogoths. They were close to reclaiming everything they lost to the Ostrogoths when the Western Empire fell (Rosen, 2007).

Even though the siege was successful in the short term, the army was weakened from the plague. The soldiers who survived were exhausted and in no state to fend off attacks. When the area was attacked by the Lombards in 568, they were defeated, and the territory was lost once again (Jacobsen, 2012).

Justinian's Reign

Justinian was the ruler of the Eastern Roman Empire during the events of 536 AD. He wanted to restore the Roman Empire to its former glory by reclaiming lost territory and proving that it was still a military threat. During his reign, he sought to regain control of Western Europe (especially Rome), revamp the legal system, and make the empire well-known once again for its culture, art, and architecture.

The Role of the Emperor or Empress

In the Byzantine Empire, the emperor or empress was a ruler with absolute power over all aspects of life, including the army, the church, government affairs, finances, and members of the council.

Both women and men could lead the Byzantine Empire. In Justinian's case, his wife Theodora ruled alongside him (Bury, 1958). She was one of the most powerful women in the history of the empire and used her power to change the policies that she cared about.

The emperor and empress were deeply tied to the church and were essentially divine figures. At the time of Justinian's reign, emperors were believed to be appointed by God (Cartwright, 2018). However, his control over all facets of life in the empire had its drawbacks.

The emperor was always at risk of being attacked, murdered, or removed from power if he made a choice that didn't suit his people. However, as long as he remained in power, the emperor would be incredibly rich and powerful.

This shows how much emperors and empresses had to lose through poor leadership. Despite Justinian's many missteps, he is still regarded as a great ruler for all of his successes.

Justinian's Successes

Legal Reform: The Justinian Code

Justinian brought about huge legal changes. He and his lawyers reviewed the entirety of Roman law to remove contradicting statements and outdated ideas. He wanted to clarify existing laws and introduce new ones where necessary. Making the laws clear and understandable was key, and Justinian and his team tried to include only one stance on each topic for that reason (Ray, 2019).

This was an important issue because past leadership had left the law in disarray.

Emperors and empresses would often add their own laws on top of the existing ones.

These often contradicted the previous laws, creating even more confusion.

Cleaning up the law would help the legal system to work more smoothly, and it also allowed Justinian to have more complete control over it. He completely changed a lot of important legislation during his time, and Justinian's revamped laws took precedence over any local laws. Plus, any laws that weren't included in his new text were repealed.

The Eastern Empire ended up using The Justinian Code until its eventual collapse almost 900 years later. By then, its influence had already spread throughout Europe. In fact, you can still find traces of the Justinian Code in both the European and American legal systems (Ray, 2019).

The Justinian Code was one of Justinian's most lasting successes, and it solidifies his place as one of the most memorable and greatest leaders of his time.

Theodora and Women's Rights

You might imagine that the Roman Empire wasn't very good about women's rights, but it wasn't as bad as you might expect. This is almost entirely thanks to the legal changes that Justinian and his wife, Theodora, made to improve conditions for women.

It's hard to determine how much these changes were based on Justinian's ideals versus how much they came from Theodora. Many scholars say that these changes were mainly Theodora's wishes, while others claim that Justinian was equally in favor.

We do know that Theodora's life prior to becoming empress was not a lavish one, and she faced a great deal of discrimination and abuse because of her gender. She grew up in a relatively low-class family, and different accounts claim that she worked as either an actress or a prostitute (Anderson & Zinsser, 1989). Regardless of Theodora's early life, she made incredible changes to the justice system for women, especially those of lower class or in unfortunate circumstances.

Under Justinian and Theodora's influence, forced prostitution became illegal, and brothels were closed. Women were allowed to own property and had more rights in the face of divorce. Theodora even made rape punishable by death.

By all accounts, Theodora was an incredible force of good when it came to women's rights and well-being in the empire. She was known to purchase girls who had been sold into prostitution, only to free them and ensure they had a good life. She even created the Metanoia, a convent for ex-prostitutes to live together and support each other. (Anderson & Zinsser, 1989)

Although Theodora wasn't technically a co-ruler, she had a dramatic influence on the laws and culture of the time period and is often considered a co-ruler in retrospect.

Taking Back What Was Lost

One of Justinian's major goals was to reclaim the territory that the Western Roman Empire lost when it fell, and he was partially successful. He reclaimed a large part of Italy, including Rome, if only for a small period of time.

While the empire still extended into North Africa and the Northwestern regions of Mesopotamia, it had lost a lot of ground in Western Europe. Justinian had a dedicated military campaign to fight back the Ostrogoths who controlled Italy and eventually reclaim that land. He reconquered most of this territory for a short while before the army was weakened by the plague.

Cultural and Architectural Developments

To restore the Roman Empire to its former glory, Justinian knew that he had to wow visitors and citizens alike with beautiful cities, incredible buildings, and an impeccable culture.

To accomplish this, Justinian invested heavily in infrastructure; fixing up old cities and roads; and creating unique and intricate buildings. One of the monuments he created that's still around today is the Hagia Sophia. The Hagia Sophia was initially built as a church but has also been a museum and is currently a mosque. This beautiful building is a testament to what people in the Middle Ages were capable of. If only we had more artifacts from this time period!

The Downsides

Justinian had numerous accomplishments and is known as a great leader, but he's done some questionable things. A lot of these could be chalked up to the extreme circumstances of 536 AD, but they're still worth noting (and critiquing).

Taxing the Dead

One of Justinian's most infamous policies was taxing the dead. Yes, he really did that!

Basically, he would tax the people who survived the plague for the amount that they owed *plus* the amount that their deceased neighbors owed. Obviously, this was incredibly frustrating for the people of the empire who were barely scraping by.

To make things even worse, these tax payments weren't going toward relief payments for the sick or even for the survivors who had been out of work due to the pandemic.

These high taxes were to cover Justinian's building projects and high military expenses— not for the people who had suffered. This seems like an incredibly insensitive move, and the people of the empire were rightfully infuriated. Contemporary historian Procopius strongly believed that Justinian's rule was unfair and inconsistent (Mark, 2020). Procopius notes that Justinian didn't impose laws or rules to try to help the sick and never gave them tax breaks or relief funds. When people discovered that staying away from others kept them from getting sick, they chose to do it for themselves—never under Justinian's guidance.

To his credit, Justinian also fell sick with the plague, although he managed to survive it. His lack of regulations or relief for the sick may be partly due to his own illness, but it still isn't a good excuse.

Justinian also may have been funding his projects instead of providing relief payments for a noble reason. He may have been trying to appease God by building churches, thinking that this might stop Him from allowing the plague to spread (Mark, 2020). This goes to show the strong influence of religion at the time, which is similar to what we saw throughout the world in response to the events of 536 AD—more on that later.

Glorifying the Empire at His People's Expense

There's also the issue of the Gothic War, and Justinian still had troops fighting in Italy throughout the pandemic. We can only speculate as to why he didn't pull his troops out during the plague. He may have thought that the disease would also weaken his enemy, allowing him to be successful. Alternatively, he might have just been so focused on his goal of restoring the empire to its former glory that he tried to push through it.

This would be in line with his focus on building churches rather than taking care of his people, too. Justinian may have thought that focusing on grand gestures, like building a beautiful Church or reconquering lost territory, would be more impactful for the empire in the long term. Since the Hagia Sophia still stands today, this may have been the truth.

While it may have been better for the empire as a whole, these were major sacrifices for the people who lived there. His citizens literally gave their lives and well-being in exchange for the empire's glory. I doubt it's a trade that his people would have made if they had a choice.

The Good and Bad of Justinian: A Conclusion

We still look back on Justinian today as an example of a great leader. He was Justinian the Great, after all. He made incredible advancements to society at the time and did a lot to improve the daily lives of his people. Clearing up the law and improving women's rights are two great examples of that.

However, he was also full of lofty dreams about restoring the Roman Empire, and these kept him from intervening properly in times of strife. He remained focused on improving the perception of the empire by reconquering lost territory and building grand structures. He focused on those goals—even while losing between one-third and one-half of his population. It's hard to say if this was a sign of his determination or negligence (or a mixture of both) when it came to his people.

We do know that the lack of modern medicine and strong religious background would have greatly influenced his decisions. You may think that you would make different choices, but we have different worldviews and knowledge than we would have had back then. At the end of the day, you have to decide for yourself if you think Justinian ruled fairly.

CHAPTER 4

THE PLAGUE

We touched briefly on Justinian's response to the plague, but how did the plague start? What would it have been like to live through a pandemic in the Middle Ages, and is it similar to what we went through in 2020? In this chapter, we'll dive into how the plague impacted different societies and how they responded to it. Finally, you'll see how the Middle Age plague experience compares with the modern pandemic experience.

FIRST ACCOUNT OF THE PLAGUE AND ITS ORIGINS

The first ever report of the plague was in 541 AD in Pelusium, a port town in Egypt (Sarris, 2021). It was an important location for trade because it connected the Roman Empire with Mesopotamia across the Red Sea. The disease spread to towns along the coast before appearing in Constantinople and the rest of the Eastern Roman Empire. From there, it went on to impact Western Europe and Mesopotamia.

The plague seemed to come to Egypt from somewhere in East Africa or Southern Arabia, but that's nowhere near where it started. It's believed that the disease originated somewhere in Central Asia hundreds of years before, as early as the Bronze Age (Sarris, 2021). It mutated multiple times over the years as it was passed along trading routes and battlefields.

It was carried by rats and fleas, which were attracted to the goods carried along trade routes and by troops. This is what happened in Egypt, too. As mentioned earlier, rats and fleas thrived on the grains being held in Egypt, and when those grains were transported, the infected pests were transported with them.

Even before the eruption of 536 AD, the Roman Empire imported a lot of their grain from other areas of the world, namely North Africa (Horgan, 2014). They also imported lots of other goods from there, including ivory, paper, and unfortunately, slave laborers. There were plenty of ways for the rats to find their way along the extensive network of trade routes.

Plus, contemporaries relied more heavily on imports as their crops failed at home. The climate changes caused by the eruption made people more reliant on trade and also may have caused infected rats to move into the area. The eruption caused a perfect storm that helped the plague spread.

Living Through the Plague

As survivors of the COVID-19 pandemic, you may feel like you already know what it would be like to live through a plague. However, the plague of 541 was different. Without our modern understanding of medicine, no vaccines, and no real idea what was causing the illness, people in 541 were completely caught off guard by the plague.

One of the most common ways to try and heal infected people in 541 AD was through prayer (Little, 2007). While this may have been comforting, it was far from a real solution.

Doctors at the time were trained in the concept of bodily humors or fluids. The theory was that if the bodily fluids weren't in balance with each other, you would get sick. To heal the sick person, you would then get rid of whichever fluid the person had too much of. This led to a healing technique called bloodletting, which is what it sounds like: the removal of blood to try and heal whatever ailment the victim suffered from. This technique, alongside other ineffective home remedies, was common (Rosen, 2007).

Symptoms

The Plague of Justinian was caused by the same bacteria that would go on to cause the Black Death in 1346. Although it took a dramatic toll on people in 541 AD, historians believe it had nowhere near the impact of the Black Death. This might be due to changes in the bacteria itself, more highly populated cities, or another factor. Regardless, the symptoms were nearly identical (Sessa, 2020). The most common symptoms were:

- A sudden fever and chills
- Fatigue or weakness
- Muscle pain
- Severely swollen lymph nodes (buboes) growing to about the size of an egg
- Blackening of the skin on the fingers or toes
- Vomiting
- Bleeding from the mouth or nose
- Cough
- Symptoms of shock (clammy skin, enlarged pupils, sudden behavioral changes)

The buboes that the plague is most known for aren't the most dangerous symptom. These were common in both mild and severe cases, and although they looked scary, they were just very swollen lymph nodes. The most alarming symptom was actually a cough, which indicated the most transmissible and quickly progressing form of the plague (Sessa, 2020).

SOCIAL UPHEAVAL

The plague outbreak changed how contemporaries lived their lives. According to Procopius (1916), it became rare to see other people outside, and if you did, they were either dying or carrying the dead.

He describes a complete social upheaval, where slaves no longer obeyed their masters, and rich people were often left to fend for themselves as their servants were either sick or had already died. This led to what he called a "universal destitution" (Procopius, 1916).

At the same time, people were either too scared or unable to work, meaning that a lot of necessary jobs like plowing the fields or harvesting crops just didn't get done.

It's important to note that this particular outbreak of the plague only lasted for approximately three to four months in Constantinople. So, while these changes are huge and catastrophic, they didn't last forever. Plus, while we have firsthand accounts of these experiences in Constantinople, things were different in other places. A farmer outside the city may have been hardly impacted by the outbreak, and people in other towns and cities weren't all hit as hard as Constantinople.

However, historical accounts suggest that the plague caused some social upheaval nearly everywhere that it struck. One account of the pandemic in North Africa says that plague survivors fought over the victim's properties, and due to financial hardship, men preferred to marry wealthy widows over younger women (Little, 2007). There are similar accounts of distress and unrest everywhere from Ireland to the Sasanian Empire.

New research suggests that these accounts might be exaggerated as contemporaries pushed their own agendas. For example, Procopius is known for hating Justinian and may have exaggerated the impact of the disease because he saw it as a reflection of his incompetence or even of God disliking him (Mordechai et al., 2019). However, the fact that we see multiple accounts with similar stories suggest that there was at least some amount of social unrest due to the pandemic.

Although the plague didn't last long in Constantinople the first time, it impacted different parts of the Mediterranean for over 200 years (Horgan, 2014). It appeared and reappeared throughout different regions as it mutated and was redistributed along trade routes before finally disappearing until the Black Death.

Personal Struggles

Regardless of whether there was mild or extreme social upheaval, living through the plague was taxing on every level. Seeing people around you die and not being able to protect them (or yourself) would be traumatic. Plus, the plague took so many lives, especially in Constantinople. Contemporaries explain that when there was finally time and manpower to bury the bodies, they had to be moved outside the city instead of being buried on burial grounds in town. They say that the smell of death permeated the whole city, which set the people even more on edge (Procopius, 1916).

At the same time, survivors of the plague were still grappling with the Mini Ice Age and the famine that came from it. Those who didn't die from the plague were still at risk of dying from starvation, especially if they were afraid to leave their homes and share food with others.

All of this happened during ongoing wars and internal conflicts, too. People in the army who survived the pandemic would have to continue fighting, even in their weakened state, not knowing how their family back home was doing.

Taken together, we can see how terrifying it would have been to live through the Plague of Justinian. Even the "lucky" survivors wouldn't emerge unscathed and struggled with all the other misfortunes that came from the eruption and the plague.

Eventual Recovery

Recovery was a slow and tumultuous process for Constantinople and the Mediterranean. Using ice core analysis, we can see when cities were in economic decline by looking for traces of mining by-products. After all, societies will only be focused on mining and refining materials when they aren't worrying about their basic survival. We can also see the type of material they were mining to see the extent of their success. For example, whether they mined lead, salt, or silver would show what the contemporaries valued and needed at the time. The more precious the metal, the more likely it was that the society was doing well.

As mentioned, the Mediterranean took repeated blows from the plague as the bacteria-carrying rats and fleas moved. The plague hit different areas at different times over a 200-year period before disappearing until the beginning of the Black Death. Economically, it seems like the region didn't truly recover until almost 100 years after in 640 AD (Gibbons, 2018). This is where we finally see an increase in the by-products for silver mining, showing that the region was recovering. It also marks the change from gold to silver coins, which brought huge positive changes including long-distance trade and bigger trading hubs (Loveluck et al., 2018).

These economic changes tell us that the region did recover, but they don't show us what it was like. What did that recovery look and feel like for contemporary people?

For the people of the Eastern Roman Empire, recovery came with some dramatic cultural and societal shifts. This is what changed the Eastern Roman Empire from essentially Roman into its own distinct empire. The first of these changes was the switch from Roman religion to Christianity.

The second was the switch in language from Latin to Greek, and this occurred slowly throughout Justinian's reign and afterward. For instance, some of Justinian's early laws are written in Latin, while later ones are written in Greek (Keum Young et al., 2017).

Similar societal shifts happened in North America as a result of the famines caused by the eruption. However, these communities recovered more quickly because of how they responded to the disaster. They moved from small family homes into larger settlements, which led to less individual work, more population growth, and better technology (Sinensky et al., 2021). These regions were lucky to not be hit by the plague until much later. Still, focusing on the community prevented them from getting malnourished as everyone provided for each other. There were more people in these larger communities to care for the sick and find food, which made it easier to sustain the whole population.

Comparisons With the COVID-19 Pandemic

As survivors of the COVID-19 pandemic, thinking about historical plagues tends to draw our own experiences to the forefront. How was the plague outbreak of 541 AD similar to our modern pandemic experience, and how was it different? There are a few key similarities and a few key differences. Please note that I'll be referring to both pandemics in past tense as we reflect on our experiences so far—not to imply that COVID-19 is over.

The Role of Geography

Geography had an important role in both the COVID-19 pandemic and the Plague of Justinian. In both cases, where you live had a huge impact on whether or not you would get sick. Living in a big city put you at a higher risk, whether you were wealthy or poor, because you had a better chance of being exposed to the disease.

If you lived on a farm in the middle of nowhere, your chances of getting sick were much lower than if you lived in the city. In either outbreak, it wasn't the only factor, but it was an important one. However, I'd argue that it was much more impactful for the 541 pandemic.

This is partially because of modern medicine and science. Since we knew early on how COVID-19 spread, we were able to create local and global policies that slowed it. These ranged from restricted travel to wearing masks and social distancing. Using these policies, living in a big city became less of a death sentence and more of an inconvenience. It meant that you would have to take more precautions when going out to protect yourself from getting sick. Similarly, people were much less likely to live completely off the grid in 2020 than in 541. People usually still had to go to stores to get things that they needed and would drive to get those things instead of going without them. We also had modern technology to do things like arranging contactless grocery pickups and checking in on our loved ones, which also prevented the spread regardless of where you lived.

In comparison, the people of 541 AD didn't know what caused the disease. They caught on that social distancing seemed to prevent it only after a huge amount of people had died. They had ineffective medical techniques and didn't have the modern conveniences we had to prevent the spread, meaning that where you live mattered a lot more.

The Role of Wealth

In places like the U.S. where healthcare is privatized, individual wealth had a huge impact on your well-being during the COVID-19 pandemic. It would change how quickly you would receive care if you got sick and how high quality that care would be. However, the wealth of the country you lived in also had a huge impact. It influenced whether or not you would be able to social distance or get protective equipment, like masks or face shields. Later on, it would influence who got vaccinated and how quickly they could do so.

Wealth also determined if you could afford to stay home and social distance during the pandemic or were forced to go to a high-risk frontline job. Some wealthy people were able to move to lower risk areas during the pandemic, which also reduced their risk of exposure.

In some ways, these factors mattered way less in 541, mostly because the healthcare you received wouldn't have helped much regardless. Whether you were wealthy or poor, you were still just as likely to get sick in a society that didn't understand how that sickness worked. Even the best doctors couldn't do much to help you recover.

However, wealth did still have some impact. If you were poor, you were more likely to have a weak immune system from lack of nourishment, especially during a famine. This would put you at greater risk of getting sick and having a worse outcome, too. If you were poor, your family and close friends were likely poor, which meant that your friends were more likely to get sick. This was a double whammy because you would be more stressed about losing your loved ones and also have less people to take care of you if you got sick.

Other Instigating Factors

For many of us, it seems like the COVID-19 pandemic came out of nowhere, but I'd argue that's far from the truth. The emergence of the virus may have been a fluke disaster, but the factors that helped it spread were not. Some of them are eerily similar to what happened in the Plague of Justinian.

An Interconnected World

It's no secret that our modern world is extremely interconnected. You can take a plane across the globe, drive across the continent, and send a text message to anywhere in the world. Global trade and travel definitely contributed to the spread of the COVID-19 pandemic, especially in its earliest phases. The same was true for the plague, which was mostly spread along trade routes. If it wasn't for these connections, the plague would have had a localized impact.

With COVID-19, these connections also protected us in some ways. It meant that vaccines created in one country could be distributed around the world. We could see the effects of the virus in other countries, and governments could learn from each other's successes and failures. This pressured countries worldwide to prevent the spread with effective policies.

Climate Change

The 541 AD plague outbreak was partially caused and spread by the global cooling that happened as a result of the eruption. We know that rat movement patterns were changed, bringing infected rats into the region.

You might not have realized that climate change had an indirect effect on the COVID-19 pandemic, too. Global warming causes more intense and frequent storms, including hurricanes, tsunamis, forest fires, and tornadoes. These kinds of storms can wreak havoc on a region, rendering much of the population homeless and restricting their access to food and clean water. Modern emergency relief plans can reduce the impact of these crises, but they're still stressful events that make it difficult to social distance and wear a mask properly.

For example, following all the proper social distancing procedures reduces the capacity of emergency shelters. In crises, this can force response teams to decide between increased risk of COVID-19 infection or protecting more people from the disaster (Centers for Disease Control and Prevention [CDC], 2021).

Similarly, global warming leads to an increase in wildfires. Getting exposed to the smoke can irritate your lungs and even reduce immune function—both of which can make you more vulnerable to COVID-19 (CDC, 2021).

This is one of the most alarming parallels between the Plague of Justinian and the COVID-19 pandemic, especially because it's one that tends to go unnoticed. If global warming continues to get worse, we'll see even more of its negative effects on our health.

Pandemics and Climate Change

Drastic climate change is frequently related to an outbreak of one kind or another. This is why, as discussed earlier, we often see pandemics right after a large volcanic eruption. The sudden cooling effect of the eruption helps the disease spread in several key ways:

- It causes changes in the movement of rats, fleas, and other animals that might carry disease, often into more populated areas where food and warmth are readily available.
- It causes crops to fail, leading to famines that weaken the population and make them vulnerable to getting sick.
- It causes people to stay in enclosed spaces, often (unknowingly) with small animals that carry the disease.

These three major factors partially explain why the Plague of Justinian happened and why it spread so quickly. However, there doesn't have to be a *sudden* cooling event to lead to a pandemic. In some circumstances, it might be (and has been historically) enough for a natural global cooling to occur—even without sudden crop failure. Infected animals would still be driven into closer contact with humans, and people would still stay in enclosed spaces for warmth.

Global Warming and the Plague

Movement of Disease-Carrying Animals

Interestingly, global warming can also cause outbreaks. This is because when the plague happens in nature, it usually impacts rodents. When it gets warmer and more humid in a certain region, we tend to see less rodents there—often because they're more mobile and moving more rapidly between regions. Fleas that carry the plague bacteria will look for other hosts, like humans and their pets (Zielinski, 2015). Plus, other pests like mosquitos tend to multiply in the warmer, wetter climates that we can expect to see moving forward. Besides just being a nuisance, mosquitos can also carry diseases and infect the people they bite.

Temperature Changes and the Body

We've seen this happen historically in natural periods of warmth, but something is different today. For one, human influences on global warming mean that we're drastically exceeding natural rates of warming. This is mainly because of carbon dioxide released into the atmosphere by human activities like using fossil fuels, deforestation, and agriculture.

In the last 60 million years, the world has heated up by four to six degrees as part of Earth's natural heating and cooling cycles (Worral, 2017). However, human influence is set to raise the temperature by another four to six degrees in only *100 years.* This is way less time for a big temperature change, and that comes with its own consequences.

Beyond the normal side effects of a warmer climate, like flooding, volcanic eruptions, and storms, we're also dealing with fragile human bodies operating in temperatures that they aren't adjusted to. Increasing the temperature by even a few degrees can have serious health consequences—all because of the stress it puts on the body (Worral, 2017).

So, in the coming warmer climates, animals that carry diseases will be more active, and our bodies will also be more susceptible to disease. That's not exactly a winning combination.

The Role of Environmental Changes

The flooding and frequent storms that come with global warming are yet another risk factor for humans. They pose a threat to cities on the coast or in the path of the storm and also lead to dramatic changes in local ecosystems. Increased heat and humidity can turn farmland into swamp. This can cause food shortages and disruptions as we scramble to produce food in new areas. In turn, these shortages can cause nutrient deficits and famines, making people more vulnerable to illness.

It doesn't help that modern climate change is driven by an excess of carbon dioxide in the air. Carbon dioxide can increase the intensity of Southern Oscillation events, where a shift in ocean currents and winds can either cause warming or cooling, depending on your location. This can cause dramatic regional temperature fluctuations, which have been historically associated with sickness (Worral, 2017).

Global warming won't mean more in terms of mild winters and balmy summers. It'll cause more intense storms; disruptions in production and food; and more vulnerability to disease. We need to step in and change the way we live before it's too late. We'll delve deeper into this in the last chapter.

CHAPTER 5

ONGOING CONFLICT

The Plague of Justinian created a huge crisis. Massive amounts of people died as a result of the combination of the plague and global climate changes. Political leaders saw this as a perfect time to strike, while opposing civilizations were weak. This led to a number of wars throughout the century and even greater losses.

THE GOTHIC WAR OF 535–554 AD

The Gothic War was between the Eastern Roman Empire and the Ostrogoths. It was one of the major wars that occurred during the events of 536 AD and the following plague. The conflict began in earnest in 536 and was fought over control of land in Italy, which used to belong to the Western Roman Empire. The chain of events that started the war began long before the chaos of 536.

THE GOTHS: A BRIEF HISTORY

Who were the Ostrogoths, and why did the Eastern Roman Empire go to war with them?

The Ostrogoths were a powerful Germanic tribe. They were a branch of a much larger Germanic group, including both the Ostrogoths and the Visigoths. Together, the two groups occupied a large part of Spain; almost half of modern France; Italy; and parts of Croatia and Slovenia at different times throughout history (Wolfram, 1990).

The Goths had a long history. In the first century, they and their neighboring tribe, the Vandals, were ruled together by another Germanic tribe. This didn't last long, and the Goths had ongoing conflicts with both the Vandals and the Roman Empire for centuries.

By the third century, the Goths had an interesting relationship with the Roman Empire. Members of the Goth tribes were often recruited by the Roman army and were heavily involved with the Roman-Persian wars (Wolfram, 1990). However, the Gothic tribes continued to have a tumultuous relationship with the Romans. They were known for their repeated and successful raids on the Roman Empire—even killing Emperor Decius in the Battle of Abritus.

The Goths continued to grow in power and carry out multiple campaigns against both the Romans and the surrounding tribes and kingdoms. However, over the centuries, they also grew closer and influenced each other's customs. Gothic clothing even became fashionable in Rome, much to the dismay of the more traditional Romans (Halsall, 2014).

The Goths also had ongoing involvement in the Roman military, frequently achieving high ranks and being known for their military prowess (Wolfram, 1990). Historians believe that without Gothic military involvement, the Roman Empire would have collapsed much earlier.

Conflict between the Romans and the Goths resurfaced when the Goths were overrun by the Huns in the late 300s, driving them to seek refuge in the Roman Empire. They were infamously mistreated by Roman officials. Starving refugees were coerced into selling their children into slavery for only dog meat in return. This, alongside ongoing civil disputes, led to the first Gothic War in 376. Ongoing tensions between leaders and the working population meant that a portion of the Romans fought alongside the Goths (Jacobsen, 2012).

In the resulting battles, Roman soldiers killed Gothic civilians and soldiers alike, only worsening the conflict as knowledge of their mistreatment spread. The Western Roman Empire had increasingly little power as its enemies grew in number and strength. Meanwhile, its own resources dwindled. The army became relatively weak without the help of the Goths—not to mention their own people fighting against them.

This eventually led to the fall of the Western Roman Empire. By 476, most historians agree that it had almost no power. The emperor was overthrown by Germanic King Odoacer, and most of the territory that had once belonged to the empire ended up in the power of Ostrogothic tribes and kingdoms (Jacobsen, 2012).

Beginnings

After the fall of the Western Roman Empire, issues between the Eastern Roman Empire and the Ostrogoths wouldn't pop up again until 535. This was mainly because it just wasn't a good time to strike. Justinian still wanted to reclaim Rome as an important symbol of the old Empire, but trade with the Goths was good, and the Gothic kingdoms had continually strengthened over the years.

The death of the king of Italy in 526 led to a period of confusion and chaos in the region, which Justinian saw as an opportunity (Ray, 2014). In 535, he declared war on the Ostrogoths.

The First Conflict

After the old Roman Empire's repeated tussles with the Ostrogoths, Justinian knew better than to just send one army. He sent an army directly West to meet the Ostrogoths head-on. Then, he sent a second army by sea to the Roman territories in North Africa, which gave them quick access to Italy. The second army was led by Justinian's most well-known and best general, Belisarius (Jacobsen, 2012).

By the end of the first year, Belisarius and his troops had conquered Sicily, but they weren't able to keep going. Issues in Roman North Africa forced him to return with his soldiers to Carthage to deal with a mutiny (Jacobsen, 2012). This ended up not making a big difference in the conflict because they were able to repress the issues at home and continue their conquest of Italy.

They met little resistance and were able to reclaim Rome. However, the general was forced to leave troops at garrisons, or small military bases, in the cities they had conquered. This meant that, alongside troops lost in minor battles along the way, his army of around 7,000 was weakened to around 5,000. The army prepared for an attack from the Ostrogoths by stockpiling food and resources and fortifying the borders of the city.

At first, the Romans were able to fend off the attacking Ostrogoths. During desperate times, they even forced the retreat of their soldiers by dropping statues on them from the walls of Rome. Then, the Ostrogoths were weakened enough that the general was able to get all the women and children out of the city. At this point, he hired the remaining men as soldiers to help strengthen their numbers (Bury, 1958).

Finally, the general received reinforcements. Researchers estimate that 1,600 horsemen were sent, which gave the Romans an advantage. The Ostrogoths weren't able to use bows and arrows from horseback, which made a huge difference in the battles (Jacobsen, 2012). The Roman troops were able to quickly approach and use long-distance attacks to devastate the Ostrogoth encampments outside the walls of Rome then quickly retreat. This meant that even though the Ostrogoths still had more soldiers than the Romans, their numbers were slowly dwindling.

This siege is one of several that occurred throughout the war and gives you an idea of just how complex and intense these battles were. This particular siege went on for about a year with frequent conflicts and skirmishes before Belisarius and his team finally drove away the Ostrogoth soldiers.

Eventually, they were able to move forward and capture cities as far north as Ravenna. The siege on Ravenna was another long and complex expedition, but the Ostrogoths eventually surrendered.

This success was cut short by a message that arrived from Justinian, offering a peace deal to the King of Ostrogothic Italy, Witigis (Wolfram, 1990). This deal was surprisingly favorable for the Goths, and Belisarius saw it as an insult. He had already had huge success in Italy and saw no reason why they needed to stop. He refused to agree to the deal—even when Gothic leaders (who respected his military prowess) offered to make him king. He ended up returning to Constantinople soon after.

Historians believe that Justinian was feeling pressured to end the war in Italy, especially with a Persian conflict looming (Wolfram, 1990). That pressure was doubled by the famine and cold, which made these conflicts more drawn out and costly. The people back home either wanted these issues resolved (which was difficult considering the circumstances) or wanted more support for them—both of which required more people. Supporting two conflicts during such an unstable period in the empire would be incredibly risky.

Infuriatingly, Belisarius would likely have reconquered Italy for the empire in only a few months' time had he not been stopped by Justinian. If he had been allowed these few extra months, Italy might have been able to escape another 12 years of war (Bury, 1958).

Second Conflict

Between 541 and 551 the Ostrogoths were strong and successful, mostly because the

Eastern Roman Empire was struggling. There were two main reasons for this:

- The Plague of Justinian took a devastating toll on the empire and dramatically reduced their numbers, weakening their armies.
- The start of the Roman-Persian war meant that those weakened armies were required in the east, so they could not easily retaliate against an attack from the west.

Together, these huge disadvantages for the empire helped the Goths advance. In 542, Ostrogoth forces had several major successes against Roman armies, taking control of the less populated regions first to eventually surround and attack larger ones, including cities like Florence (Bury, 1958).

Their success was also due to the new King of the Ostrogoths, Totila. Unlike the Romans, Totila was surprisingly generous to the regions he conquered. When he took captives in a city, he would encourage peaceful surrender and even feed starving citizens to ensure they were in good health (Bury, 1958). These tactics encouraged people to follow him, bolstering his armies before he attempted an attack on Rome. Totila even tried to make peace with Justinian before invading Rome, but this was refused.

The second conflict began in full force when the Romans and the Persians made a five-year truce in 544, giving general Belisarius time to return to Italy. However, the aftereffects of the plague and famine left the empire starved for supplies, and Belisarius was unable to provide support to Rome until after it had been looted and partially destroyed by Totila.

Belisarius was finally able to claim Rome once again only a few months later, but he couldn't undo the damage that the walls had taken. At this point, he was called back from Italy (Bury, 1958).

When Totila attacked next, he was fought off by the city's defense, only to succeed two years later in 549. He killed everyone in the city except the women, including those who tried to escape.

That wasn't the end of it, though. In 550, the empire tried once again to reconquer Rome. This time, the Ostrogoths were seriously outnumbered. Totila tried to fool the empire's general Narses into negotiating with him while launching a secret attack on the army, but Narses was prepared and was able to decisively defeat the Ostrogoths. King Totila was killed alongside the rest of his army.

Although this marked the end of the official conflict, other Germanic tribes still had control over a large part of the region. Narses sought to defeat them to regain total control of the area, but this ended up taking longer than anticipated. He didn't finish that mission until around 562. Regardless, Italy was finally under Roman control, and the Gothic tribes in Italy were able to live peacefully there.

Throughout the entire second conflict, the plague was spreading rapidly throughout the Mediterranean. The battles of the Gothic war sped up its spread and changed the course of the war. It severely weakened the Roman army, especially with the disastrous effects that the plague had in Constantinople. This left them with fewer troops, and those that were left were weaker both physically and mentally. The plague took the lives of many, and a lot of the soldiers carried those losses with them. The plague was a huge deciding factor in the outcome of this war (Bury, 1958).

Social Upheaval and the Holy Roman Empire

The Gothic War and the ongoing crises at home weighed heavily on the people of the empire. It marked a time of transition from the Pagan Roman Empire to the Holy Roman Empire.

The Power of the Church in the Face of Disaster

The disasters starting in 536 AD played a big role in the religious transition of the empire. While it was likely to happen eventually, what seemed like the end of the world definitely increased the people's belief in Christianity.

The Church's Helper Role

Contemporaries were living through multiple disasters that made it hard to do basic jobs, including caring for the crops. This effect was multiplied by the cold, which made crops fail—even in the best of circumstances.

In these times, people often didn't have the money or the means to get food, and starvation was incredibly common. Many people relied on the church for their food.

The church's role as a helper in the community made people more likely to see their religion in a positive light. When people were dealing with serious personal losses or illness, feeling cared for by the church could have a dramatic impact on their lives.

Fear and Christianity

As mentioned, it would've been terrifying to live through such a tumultuous time. The uncertainty, fear, and distress that citizens felt needed to be relieved somehow. For many people, joining the church was a way to do that. It allowed them to feel comforted in the face of possible death because they believed they could atone for their sins and end up in heaven. This, and other Christian teachings, gave people comfort and helped them manage their circumstances.

Being involved with the church also had some practical benefits. For one, it provided a supportive community. Members of the church could rely on each other and may have even seen it as their duty to help each other (Keum Young et al., 2017). By itself, having a supportive group of people to share your troubles with is a huge advantage. It can reduce stress and make hard times less physically taxing.

It also meant more people to care for you if you got sick, which could be an incredible benefit for people who lived alone and otherwise didn't have much support.

Justinian's Role

Justinian also changed a lot of policies around this time. Many of his legal changes mention Christianity or Christians directly, giving them extra privileges. These changes were often blatantly discriminatory to other groups. For example, one of his laws required that Jewish people could not hold Christian slaves (Keum Young et al., 2017). He also would not allow a non-Christian person to be a court witness.

His most obvious and all-encompassing religious law states that anything forbidden in the Bible was also forbidden by the law (Keum Young et al., 2017). These changes reflect the dramatic religious transformations happening inside the empire. The previously pagan state began forbidding pagan rituals and made the Bible their law.

These changes also reflect Justinian's own beliefs and desires. He was Christian himself and wanted his people to perceive him as a sacred religious figure. He was quoted saying, "For what is greater or more sacred than the Imperial Majesty?" (Keum Young et al., 2017). He believed that the emperor's word should be seen as a reflection of God's word and was very closely connected with the church.

Justinian's favoring of Christianity was a dangerous beginning and one that led to the prosecution and murder of people from all other groups under the Holy Roman Empire.

On a lighter note, Justinian's new policies were also strangely at odds with the old Empire that he sought to restore. While he desperately wanted to reconquer Rome and expand the empire to the West again, he also wanted to reinstate it with his own religious values.

This was when Justinian got involved with multiple church building projects. These may have been a way to appease God and stop the plague, but they also were a way to glorify himself as a sacred leader. Many of his churches involved artwork that put him at the center of the church, showing that he wanted to be seen as the leader of the church as well as the leader of the empire. In his mind, these things went hand in hand. His church art has even been considered propaganda (Keum Young et al., 2017).

Justinian favoring Christianity would have definitely had an impact on his citizens' decisions. Being involved in other religious groups went from the norm to a reason for prosecution, and many people may have gotten involved with the church simply to protect themselves from Justinian's harsh policies against other groups.

Holy Beginnings

Although the Holy Roman Empire wouldn't come to exist officially until the 800s, these events certainly set the stage. Forbidding other religious rituals, favoring Christianity, and a close connection between the emperor and the church were all sure signs of what was to come for the empire.

From Ancient to Medieval

Alongside the religious changes, the Gothic Wars marked another key change for the Roman Empire. The empire was continuously evolving to be less and less like the old Western Roman Empire and more of its own unique blend of traditions, religion, and culture.

This was just the beginning of the transitionary period for Europe that was the Middle Ages. The events of 536 AD rocked the worlds of contemporary people, ushering in a period of dramatic change.

The Birth of Islam

The drastic climate changes of the sixth century helped Islam become an established religion. They provided huge benefits to the Arab Peninsula that allowed it to flourish, expand, and change.

The Marib Dam and Migration

Remember the Marib Dam from Chapter 2? This dam was the reason that people could grow crops in a desert. It was located on the Arab Peninsula—a region that was relatively well-off during the long period of cold and famine that Europe endured.

The Marib Dam was an incredible structure, especially for the time period. It used the fancy hydraulic technology that the local Sabeans were known for, making it possible to grow important crops like the trees that produce frankincense and myrrh. These trees were a main source of the region's wealth for many years (Paowary, 2018).

The rise of Christianity eventually reduced sales of frankincense and myrrh, which were banned for their use during pagan ceremonies. The local economy was heavily reliant on these products, and it began to suffer. The dam stopped receiving proper maintenance and eventually collapsed altogether.

The collapse of the dam meant that crops could no longer grow in the area, and this brought the finishing blow to the city of Marib. Approximately 50,000 people who lived there were forced to relocate. Locals moved permanently to more inhabitable areas of the Arab peninsula and surrounding regions.

PROPHET MUHAMMAD

Muhammad was born shortly after the collapse of the Marib Dam and is an important Islamic figure. He marked the transition to Islam and was responsible for the Muslim conquests that occurred in the region and beyond, changing the course of history.

Muhammad was born in Makkah (modernly known as Mecca) and was raised by his uncle after his father and mother passed away. He was said to have his first revelation in a cave, where he was told by God to stop people from worshiping idols. Idol worship was a common practice in the region at the time, but it was controversial enough that some people began to follow Muhammad (Yalman, 2001). This was the beginning of Islam.

When new Muslims were being persecuted in the region, he and his followers were invited to Medina. This migration was known as the Hijrah, or Hegira. There, Muhammad was able to form a new constitution and create the first Islamic state. He was eventually able to spread his reach back to Mecca and beyond, converting almost the entire Arab Peninsula to Islam.

It wasn't all peaceful, though. Many of his successes were hard-fought, coming only after all-out battles with dissenting cities and towns. Muhammad and his followers were also the targets of several assassination attempts.

The conversion of the Arab Peninsula to Islam marked a dramatic shift in religiosity in the region. Even when Muhammad died in 632, his followers were determined to spread his message (and therefore Islam) through the surrounding regions.

MUHAMMAD'S LEGACY: THE MUSLIM CONQUEST

Muhammad's followers successfully spread his ideals, constructing armies that fought and conquered other lands for Islam. They were incredibly successful, conquering parts of the Eastern Roman Empire including Syria, Egypt, and Palestine. They were able to completely take over Iraq and Iran, effectively ending the Sasanian empire. The entire Arab Peninsula, the Middle East, North Africa, and even Spain and Portugal were under Muslim control at some point in history (Lapidus, 2014). This will help you understand just how huge this empire became and how wildly successful it was.

In less than 100 years, Islam went from being isolated to parts of the Arab Peninsula to being a global religion. This dramatic change brought with it a culture and political shift as the Muslims introduced their own method of rule, including new taxes and ruling systems.

These changes were surprisingly favorable for people in the newly conquered lands compared to other leaders' policies at the time. Many people welcomed the change in leadership because it led to lower taxes and greater freedoms than they had under the previous leaders, especially areas controlled by the Byzantine Empire (Lapidus, 2014).

The Muslim conquest, unlike the Sasanian or Byzantine Empires, didn't try to subject everyone under their rule to one official religion. Instead, they allowed other religions to coexist and even cooperated with Christian religious leaders to restore churches and provide support. Despite being protected and having adequate support, non-Islamic religions were still seen as socially inferior (Lapidus, 2014).

The change of leadership didn't initially result in huge political changes. One of the policies that the leaders of the conquest followed was to keep the original leadership intact and simply control taxation (Lapidus, 2014). This helped to prevent unrest in the early years. However, the conquerors gradually increased their control, creating their own regime which the local systems and leadership would eventually serve.

Despite the huge success of these conquests, they didn't lead to a dramatic conversion to Islam—at least not directly following a conquest. These conversions didn't happen en masse until centuries later when Islam became associated with greater social status, and other religions were associated with lower status. Converting to Islam allowed nobility to maintain their high social status and was desirable for that reason (Tramontana, 2013).

Fall of the Sasanian Empire

The Sasanian Empire had been in serious decline for years, weakened by ongoing conflicts with the Eastern Roman Empire. After that, a four-year civil war left the Sasanian Empire in ruins.

When the Muslim conquerors arrived only a year after, the Sasanians put up a surprisingly strong fight. That being said, there was simply not much that they could do without an emperor and with a severely weakened army. After years of chaos and war, the Sasanian Empire finally collapsed, and the area was conquered by the Muslims in 654 (Lapidus, 2014).

Connection With the Events of 536 AD

Even hundreds of years after 536 AD, we're still seeing the fallout from the eruption. Empires that were once hugely powerful were weakened by the famine and especially by the plague.

We can't understate the importance of the sudden temperature changes that led to the death of crops across Eurasia. The resulting famine left populations vulnerable to illness— just in time for the plague to sweep through. The plague had a devastating toll in Constantinople, which was a crucial location for sea trade. From there, trade with neighboring empires, kingdoms, and tribes allowed the plague to spread even further.

In 627, the plague hit the Sasanian Empire and was a huge reason for its decline and eventual downfall. It was called the Plague of Sheroe, and around half its population died, including the king whom the plague is named after.

At this point, it's clear that the eruption led to a chain of events which completely changed the course of history, making 536 AD the worst year and the start of a miserable time. The famine and resulting plagues led to an incredible loss of human life. The people who did survive emerged weak and spent from caring for themselves and their loved ones, alongside grieving their losses. It was truly an awful time to be alive.

CHAPTER 6

THE AMERICAS

Eurasia was not the only area of the world that was devastated by the effects of the volcanic eruption in 536. North, Central, and South America were also impacted by the climate changes that caused famine, droughts, and general unrest.

NORTH AMERICAN PREHISTORY

In the 500s and the years following, populations in North and South America were booming. People were developing more permanent settlements, and as a result, they were getting more specialized skills and lifestyles that matched their environments.

Settlements in the southern regions of North America became focused on farming. In the eastern regions, developments near rivers flourished into what would later become Mississippian culture. Even Arctic tribes in modern Northern Canada were flourishing through whale hunting. Many of these areas had been populated for a long time, but the people were increasingly becoming experts in their territory. The Pre-Puebloan cultures in the Southwest had been around since about 1000 BC, for example (Metropolitan Museum of Art, 2021).

The earliest evidence we have of North American villages is also in the Southwest, where crops like squash and corn were grown. These Southwestern natives were influenced by their more advanced southern neighbors in Mexico, who grew similar crops and had a tradition of mound building.

Mound building was a custom in Mexico that was adapted by the Southwestern tribes, where huge burial mounds were built around tombs. This tradition was also adopted by the Hopewell tribes.

The Hopewell Tribes

The Hopewell tribes were a network of tribes that coexisted from about 100 BC until the 500s. They had a developed trade network and often exchanged materials with each other. These materials were used to build products, which would then be resold to other tribes on the network. These goods were often used in burials.

They tended to live in temporary settlements and moved frequently within their region. Most of these tribes were concentrated in what we would now call the Northeastern to Midwestern US, but some extended north into Canada or south to the shore of the Gulf of Mexico.

It's important to remember that these regions had highly developed skills, trading networks, and cultures—even as far back as the 500s. Although it seems like a long time ago, these were not caveman times!

South American Prehistory

South American tribes were flourishing as early as 16,500 BC. That's over 18,500 years ago! Historians believe that these settlers originally came over a land bridge that used to exist between Asia and modern Alaska, spreading from there across the continent (Johnson, 2012).

We have evidence that the South American tribes were domesticating animals like llamas and alpacas around 3500 BC, using them for transportation, meat, and shearing them to make clothing. Many regional tribes were highly involved with farming and fishing. New evidence shows that they even lived in large numbers in the Amazon rainforest, which we previously believed could not sustain many people due to the poor soil for farming (Roosevelt et al., 1996).

South America had several highly advanced civilizations. Two of the most specialized were the Inca and the Muisca.

The Inca

The Inca ruled over a huge area of land, encompassing almost 100 distinct cultural groups. They're known for their incredible stonework and metalwork. They would go on to have a developed road system and even perform highly effective skull surgeries (Norris, 2008). Much later, they also built Machu Picchu!

The Muisca

The Muisca were known by other tribes as the "Salt people" for their involvement in salt mining, most of which was done by women (Miller, 2016). They had advanced farming techniques and trade routes but were most known by the Spanish conquistadors (and perhaps still today) for their gold-working. A huge percentage of the Spanish soldiers on this conquest died on their journey to the "city of gold."

Teotihuacan

Teotihuacan was a city located in modern Mexico. It was an important religious area for the local people and grew to become the largest city in American prehistory—both in area and in population. A variety of cultural artifacts were found there, leading historians to believe it was a multicultural city. It had a huge influence on culture in the surrounding region. They undertook huge architectural projects, including the *Pyramid of the Sun* and *Pyramid of the Moon*. I can't overstate how large this city was, especially for its time period. At its largest point, it had around 125,000 citizens—about half the population of the entire Valley of Mexico at the time (Sanders & Webster, 1988).

Societies in the Americas were developing quickly and would go on to be incredibly successful. So how did the events of 536 AD affect them—if at all? Were they far enough away to avoid the brunt of the disaster?

536 AD: A Turning Point

Initially, it might seem like the eruption of 536 AD didn't have much impact on the people of the Americas. After all, they continued to thrive in the coming centuries, and they didn't have to worry about the same plague outbreak that struck in Eurasia. Maybe they were far enough away to avoid the clouds of volcanic ash that led to colder temperatures and crop failure.

Unfortunately, that's not quite what happened. Indigenous Americans were still impacted by the volcanic eruption in Iceland, shown in tree ring analysis and evidence of crop failure just after the eruption in 536 AD (Sinensky et al., 2021). We saw a resurgence of the same effects in 541, after the second eruption. This means that the Mini Ice Age wasn't just limited to Europe and Asia. It proves that the cooling was a global phenomenon and that it was intense enough to cause crop damage as far as South America.

So, while the Indigenous American people weren't dealing with a plague, they *were* dealing with a sudden temperature drop and failed crops that led to famine throughout the regions. Historians think that crop failure accelerated political changes in the region by heightening civil unrest and necessitating migration.

Teotihuacan: A Case Study

To delve into the civil unrest that happened as a result of the famine, we'll need to get a little more background on how Teotihuacan society worked. By the 500s, local society was divided into two major groups (Manzanilla, 2015). The "exclusive" group was at the top— the neighborhoods where the elite lived and made money through personal connections and long-distance trade. Meanwhile, the "corporate" group was at the bottom, which relied on menial labor to make their money. There was a huge difference in wealth between the upper and lower classes, which was a big source of tension even before the events of 536 AD.

When the temperature dropped and crops were killed, it was a much bigger blow to the poorer corporate group than the exclusive group. Corporate people were dying from starvation.

The famine had a much bigger impact on the lower class because they relied much more heavily on the success of their local crops for nourishment. For one, the poorer class didn't have the resources to stockpile food or save for later. They relied on what they could get at the moment. The elites, on the other hand, likely had large stockpiles of food and were hardly affected by the famine.

The city also had another problem: disease. Just like in Europe, the agricultural issues and famine weakened people's immune systems. They were more vulnerable to disease, and the illnesses that were common at the time reduced people's ability to absorb nutrients (Keys, 2000). This means that even when there was food, the nutrients from it weren't being properly absorbed.

Death rates were skyrocketing. About 68% of the corporate population was dying before the age of 25, which put an even bigger strain on the lower class (Keys, 2000). Previously, the rate of death for working people before age 25 was about 39%. They were losing their young people at alarming rates, but they also had to deal with the practical implications of those losses. Less young, poor people meant that there were less workers, which only exacerbated the agricultural issues. This meant even further economic collapse, which sent the city into a downward spiral.

At this point, we have proof that sculptures and statues were destroyed, important buildings were burned where the elites lived, and members of the elite were murdered (Manzanilla, 2015). It didn't stop there, though. Nearly every elite building was looted, destroyed, and then set on fire (Keys, 2000). Historians were hesitant to believe that the destruction was the result of civil unrest because it was so far-reaching. Most of the city was destroyed, which wouldn't exactly be beneficial to the lower class. However, we have no evidence that there was an invasion by some foreign enemy. This suggests that the burned buildings, shattered sculptures, and murders were all the result of a riot against the ruling powers.

The dramatic and extensive destruction of the city is an important symbol of the frustrations with the corporate class. They were so angry about the state of the city, the unfair distribution of wealth, and their poor living conditions that they were ready to burn it all down… literally. They must have known that their actions would probably have terrible consequences, leading either to the destruction of their city or their own imprisonment, banishment, or death. These people must have been desperate for any chance of change—even if it meant losing everything.

We can't say just how much the drought and famine directly caused these events. Perhaps tensions were so high that these riots were bound to happen anyway. We do have proof that there was tension between the elite and the corporate classes and a huge disparity between their wealth and well-being—even before the events of 536 AD. However, the city's agricultural difficulties had far-reaching effects, creating huge economic issues and worsening the spread of disease and infection. These effects probably sped up the chain of events that unfolded and may have led to a more dramatic outcome.

If it hadn't been for the famine, the city might have existed in the same state for a year, or 10 years, or even 100 years. Regardless, we can assert that the famine worsened the existing civil unrest and played some part in the fall of Teotihuacan.

Alarming Parallels With Modern Times

It isn't hard to see the parallel between Teotihuacan in 536 AD and our modern society. We have an unequal distribution of wealth that creates tension between the upper and lower classes—not to mention the tension between the political left and right about what should be done.

The COVID-19 pandemic only exacerbated these differences. Alongside the labor shortage, we've seen a decrease in the food and other goods that are available and spikes in cost for everything from milk to computer parts. If we were to go through a global disaster on par with 536 AD, what would happen to our society? It could only worsen the high tensions we're already dealing with.

Unfortunately, with the current state of our climate, events like this are becoming increasingly inevitable. While volcanic eruptions are common and happen naturally at all temperatures, we know that they become larger, more devastating, and more frequent when the climate warms. What about the other results of global warming that we talked about in Chapter 4?

The events of 536 AD should remind us of our fragile state as a society and what we need to do to protect our planet and ourselves. Most importantly, we need to work together to reduce our impact on the climate.

Other Settlements in the Americas

Historians have studied the effects of the 536 AD eruption on the Pueblo people extensively. As a refresher, the Pueblo lived in the Southwestern U.S. region and relied heavily on farming.

Interestingly, prior to the eruption, agriculture wasn't adopted in the same way by all tribes and villages. Different groups in the region had different agricultural strategies and may have relied more on fishing or gathering than agriculture. Agriculture wasn't equally developed in all areas either, and some tribes had more advanced farming techniques than others (Sinensky et al., 2021). This pattern went on for over 2,000 years. It's evidence that these tribes weren't in contact with each other or had very little contact.

By 580, though, it shifted drastically. Unrelated farms and tribes began to use similar agricultural techniques, technologies, and even adopted similar social and political structures (Sinensky et al., 2021). Researchers believe that this change was caused by the disaster in 536 AD.

Usually, the negative effects of temperature shifts or poor farming conditions were mitigated in these regions with social connection. Strong social networks protected families because they could communicate with their friends to see which regions were having successful harvests and move between regions as necessary. Throughout history, we see a pattern where farmers moved between areas of high and low elevation to find success.

However, this didn't work in 536 AD. We know from tree ring and ice core analysis that the most dramatic temperature shifts in the region were caused by the volcanic eruptions of 536 and 541 (Sinensky et al., 2021). In order to survive the sudden and drastic temperature changes, farmers had to move much further than they normally would, and their social networks couldn't help them enough. This caused widespread migration and the creation of larger villages. Since farmers could no longer support themselves with just their household and a small social network, larger villages helped them to deal with the changes.

These larger villages had other benefits for the farmers, leading to technological advancements and more complex social structures, meaning that they were more able to support each other throughout difficult times (Sinensky et al., 2021).

However, these developments also marked the beginning of social inequality. Farmers were forced to move into other social networks, in which they knew less people and had lower social status. At the same time, these social systems were becoming more complex with more defined ranks. These changes were beneficial because they helped contemporaries to survive their circumstances, but they also had some unfavorable consequences.

The Worst Year Ever... ?

There's no question that the events of 536 AD were hugely impactful on the Indigenous American people. The cold and famine turned existing social systems upside down and wreaked havoc on people's well-being, especially in Teotihuacan.

But was 536 AD the worst year ever for the Americas? I would argue that it wasn't. It was definitely a tumultuous year that caused a huge amount of social change, but it wasn't disastrous to the same extent as what we saw in Europe and Asia, where cold and famine led to a terrible plague, multiple large-scale conflicts, and the fall of several empires.

Instead, the "worst year ever" in the Americas would probably be around the 1520s, when Spanish conquistadors arrived and destroyed the Indigenous peoples' ways of life, infected them with smallpox and other diseases from Europe, and effectively killed off up to 90% of the population

(Diamond, 2019). That's a *huge* amount of people to lose and would have been absolutely devastating.

So, while 536 AD was the worst year in Europe and Asia, that wasn't necessarily the case in North and South America. It's important to note that the disaster impacted all areas of the world differently. The intensity of any global event depends on where you are, your local social systems, and the tensions that already exist there.

CHAPTER 7

GREAT BRITAIN AND SCANDINAVIA

The aftereffects of the volcanic eruption and resulting famine had an incredible impact on every part of the world. Regional conflicts broke out in Britain and were decided by the effects of the plague. Meanwhile, the famine caused incredible suffering, unrest, and uncertainty in Scandinavia that impacted every aspect of life, including mythology.

GREAT BRITAIN

The sixth century was a time of ongoing conflict for the British. This conflict was mainly between the Anglo-Saxons and the Romano-Britons.

Before we dive into the effects of the 536 AD eruption on Britain, it's important to note that we have very little concrete evidence about what happened during this time in Britain, especially prior to the eruption. Many of the accounts we have are from hundreds of years after the fact or from unreliable, biased sources. We simply don't have enough information to say for certain what happened. In this chapter, I'll be sharing the story that is most backed up by the research.

THE ANGLO-SAXONS

The Anglo-Saxons were the contemporary version of English people, but they weren't native to Britain. They lived along the coast of what would now be France, Belgium, the Netherlands, and Germany, bordering the North Sea.

They also didn't speak English in the way it's spoken today. We use less than 30% of their words in modern English, but the words we do share tend to be common, everyday terms (Williams, 1975).

They eventually migrated into Britain, where the Anglo-Saxon identity developed into something more similar to our modern understanding.

THE ROMANO-BRITONS

The Romano-Britons were a cultural group that lived throughout Britain until the sixth century. Until the early 400s, Britain was part of the Roman Empire, and Romano-Britons tended to adopt Roman culture and religious beliefs. The Romano-Britons are the ancestors of modern-day Welsh people.

CONFLICT IN SIXTH-CENTURY GREAT BRITAIN

In the 300s, Britain and the surrounding regions were relatively peaceful despite the migrations that were happening. This was mainly due to the Roman troops who protected the region from external threats and dealt with any internal conflicts.

These migrations involved Germanic tribes extending toward Britain—the same way they had pushed into Italy. We know that the Western Roman Empire wasn't exactly thriving at this point. They were forced to call their troops out of Britain and into other regions where they were needed more. Namely, this was to deal with the Huns who were invading in the east.

However, the Roman influence on British culture didn't just disappear. They were still involved in trade and were technically part of the empire for some time. Roman culture stuck around—even when the Romans didn't.

The British were suddenly fending for themselves, and this was when the conflict began in earnest. When the Romans withdrew from Britain, it left them vulnerable at a crucial time. They were already dealing with invasions and attacks from the Picts and Scots. The Picts were a Celtic tribe that lived in Scotland whose name in Latin means "painted ones." They got their name for their elaborate whole-body paint, which may have even been tattoos. The Scots, on the other hand, came from the combination of the Picts and the Gaels.

Together, these two groups posed a big threat to the British. Since the Roman emperor refused to send them help, the rulers decided to take things into their own hands.

By then, Britain had fallen into the hands of local warlords. Some contemporary sources claim that a specific warlord named Vortigern, or the "proud tyrant," became king at this time (Monmouth, 1136/1999). In order to deal with the invasions, he decided to hire Anglo-Saxon troops to strengthen their borders against the threat from the northeast. This was in addition to the Germanic soldiers that had already been working there.

The Germanic and Anglo-Saxon troops were able to turn the tides of the fight against the Picts and the Scots. However, victory didn't last long. The soldiers quickly became dissatisfied when they felt that the rights of their contracts weren't being held up, and they were being denied payment in the form of rationed food. They began to revolt against the Romano-Britons, slowly claiming more and more territory.

As people continued to migrate into Britain from the east, particularly Anglo-Saxons and Germanic people, they may have joined the battle and eventually caused the British to lose ground.

However, the British had several major successes. One of these was the Battle of Badon,

a relatively well-known battle around 500 AD that prevented the Anglo-Saxons from advancing against the Britons for some time.

Following the battle, there was a period of relative peace. During this time, the Anglo-Saxons established communities in Britain's northern regions, known at the time as Northumbria. These communities grew into larger areas of Anglo-Saxon control with their own kings.

Of course, peace couldn't last for long. The migrating Anglo-Saxons and Germanic people continued to grow in number, worsening tensions until another battle occurred. This one was called the Battle of Bedcanford and was decisively won by the Anglo-Saxons in 571. They were able to gain control of four more settlements, greatly expanding their territory (Monmouth, 1136/1999).

From there, the Anglo-Saxon kingdoms grew even further and became well-established. Tensions were still high between the Anglo-Saxons and the Britons, and various battles happened between them from the fifth century into the eleventh century, increasingly in favor of the Anglo-Saxons.

Over time, evidence of the original Romano-British people decreased. Some historians believe that they were pushed into the mountainous Western regions of Britain or continued in small communities interspersed in the territory. Others believe that the Romano-British were essentially assimilated into the Anglo-Saxon communities and culture. All of those things probably happened to some extent. Given the location of modern Wales, it would make sense if many of the Romano-British people migrated into this area because they were the precursor to the Welsh people. However, a good portion of the population was probably assimilated into the Anglo-Saxon kingdoms, especially as they conquered preexisting Romano-British towns and settlements.

The Impact of the Icelandic Eruption

What did the Icelandic eruption have to do with all this? More than you probably think!

When the eruption occurred in 536, it caused widespread famine in Britain and Ireland as it had around the globe. The Gaelic people were the ones who reported the infamous "failure of bread" from 536–539. It's important to keep those numbers in mind. This was a long-term crop failure for not just one season but *three years*. A famine of this length on the population would have had extremely widespread and devastating effects.

The region was highly reliant on farming for most of their nutrients, and crop failure led to starvation and nutrient deficiencies, which—you guessed it—made the people's immune systems weak. Unsurprisingly, it also took a toll on their armies. This meant that during the conflicts in Britain, soldiers were often fighting through starvation.

These conflicts would have also taken their toll on the rest of the population—ongoing instability increased their stress on top of what must have felt like a never-ending famine.

The famine was incredibly widespread and took its toll on both the Romano-British and the Anglo-Saxon people (Keys, 2000). However, historians believe that the continued migration of the Germanic tribes into Britain was hugely beneficial to the Anglo-Saxons. While Romano-Briton numbers were dropping to the famine, more Germanic and Anglo-Saxon migrants were arriving from across the sea. So, even though they were still experiencing huge losses, they were somewhat mitigated by the migrants.

After the famine, the plague struck Britain in 549, probably as a result of trade with the Mediterranean. There was huge loss of life there which dramatically reduced the population of many of the neighboring tribes, alongside the Romano-Britons and the Anglo-Saxons. We have proof of these losses through records of complete depopulation of areas that were previously successful settlements, right around the time of the plague. An estimated 65% of the population died in southwest Britain (Keys, 2000).

Just like the famine, this was devastating for both sides of the conflict but especially for the British. While the British dealt with huge losses, the Anglo-Saxon forces continued to be fueled by newcomers from overseas. This is one major reason for the huge Anglo-Saxon successes during this time. They had a steady stream of new soldiers to replace those lost to the aftereffects of the eruption while the British didn't.

The dramatic climate changes also caused big storms with many casualties. In the years after the eruption, contemporaries in Britain describe frequent storms, huge hail, and very cold winters that killed hundreds (Keys, 2000).

This provides important context for the conflicts in Britain during this time. While the conflicts were going on, there was a famine, an increase in storms, and a raging plague. This partially explains why there is so little evidence of what happened during these battles: Many texts would have been lost in disasters like these. Plus, the people who survived were focused on their ongoing struggle—not on preserving documents. Those on the front lines may have been blessed with whatever food rations were available, but they still faced a pandemic and the loss of their friends and loved ones—to the conflict as well as famine and disease.

King Arthur: A Real-Life Hero?

Most of us are familiar with the story of King Arthur. According to some sources, he was the leader responsible for defending Romano-Briton against the Anglo-Saxon invasions during this time. Is this really true, though?

These beliefs are based on texts by two writers in the ninth and tenth centuries, long after King Arthur died (History, 2018). That knowledge alone raises some questions. Most importantly, why did such an important historical figure go unmentioned until hundreds of years after his death?

There are a few possible solutions. Maybe the story of King Arthur was so well-known in oral traditions that no one felt the need to write it down until later. This has happened with historical figures before, where knowledge of them was so widespread that contemporary writers simply expected others to know the story. Or, since we have so few surviving documents from sixth-century Britain, we might just be missing the texts that referenced him. Either way, this would mean that King Arthur was real; we just don't have a lot of evidence of him.

Alternatively, King Arthur may have never existed. He could have just been a mythical hero who was attributed to leading the Romano-British. This would mean that there was another person who actually did these things but who didn't necessarily line up with the mythical figure of King Arthur. Some evidence does point to this idea. Welsh historian Nennius wrote about battles that King Arthur supposedly participated in, but they occurred in so many different locations and time periods that it would have been physically impossible for one man to be involved in all of them (History, 2018). It's possible that King Arthur was made up to encompass all the generals and kings who fought in these important battles.

This is the perspective that most historians believe in today (Higham, 2021). With only feeble, contradictory evidence, it's hard to make a case for King Arthur's existence. So, although the legend says that King Arthur fought for the Romano-Britons and died in the Battle of Camlann, we have no reason to believe he was a real historical person.

Still, the King Arthur legend tells us how important these battles were to the history of Britain, especially to the Romano-British people, and gives us new insight into their perception of history. As they were fighting a losing battle against the Anglo-Saxons, King Arthur stepped in and saved them. It's a beautiful story of people creating the hero that they needed.

Scandinavia

We learned about how the aftereffects of the eruption devastated Europe, the Mediterranean, and even impacted the Americas. But how did Scandinavia fare in the fallout of the Icelandic eruption?

Too Close to Home

Unfortunately, Scandinavia was one of the locations that was hit the hardest by the catastrophe in Iceland, probably because they're so close geographically. Much like the rest of the world, we were able to see how suddenly and drastically the climate changed using tree ring analysis. Analyses in the region showed dramatic reductions in tree growth between 536 and 542, and they didn't fully recover until the 550s (Keys, 2000). This is proof that a long cold snap also impacted Scandinavia.

As in the rest of the world, the cold had huge impacts on the landscape and agriculture. Famine ran rampant, accompanied by increased rainfall and humidity. Lack of evaporation caused by the volcanic ash "dust veil" worsened these effects, causing water levels to rise and rivers to overflow (Maraschi, 2019). As a result of the famine and drastic climate changes, 75% of villages were abandoned in parts of Sweden (Gräslund & Price, 2012). This huge depopulation shows just how devastating these climate changes were for the Scandinavian people.

On top of that, there was up to a 90% decrease in proper burial services (Gräslund & Price, 2012). Only the rare few were lucky enough to receive a proper burial, further highlighting just how grave the circumstances were in Scandinavia. People were so poor and weak, and dying so frequently, that they were unable to carry out even the most basic rituals for their loved ones.

It would have been terrifying to live through a disaster like this. Scandinavia saw drastic drops in summer temperatures, which meant that there was no relief from the constant cold. Food scarcity made a bad situation completely dire. Those who did survive were extremely lucky and often narrowly escaped death themselves.

Mythological, Religious, and Cultural Effects

The cold, weak sunlight and failing crops also had huge cultural and spiritual influences. Just like in Europe and the Mediterranean, the aftermath of the eruption seemed like a bad omen for the end of the world. Without modern science, most people assumed that the events had religious connotations. Well, the same was likely true in Scandinavia.

The Scandinavian people had a religious tradition of offering their valuables, particularly gold, as sacrifices to the gods. This was a way of appeasing the gods to increase the chances of having their wishes fulfilled.

Interestingly, huge amounts of gold have been found which date approximately to the period after the eruption. Over 25 pounds of gold was found at just *one* of these sites, and there were many more (Axboe, 1999). While Scandinavian settlements were well developed, they weren't huge cities. These were towns and villages parting with all of their valuables in order to appease the gods.

The huge increase in gold sacrifices immediately after the 536 event is no coincidence. Contemporaries were using what they knew, including their religious beliefs, to try and bring back the sun and end the ongoing winter (Axboe, 1999).

Some of these sacrifices were what archeologists call official offerings. These were made by the ruling class on behalf of the community, showing an organized, desperate attempt to change their circumstances.

Many other valuable items were sacrificed besides gold, including other types of jewelry or objects of high social status. These items were treasured by their owners and were often buried with them or passed along as heirlooms in normal times.

These gold sacrifices give us new insight into what it would have felt like to live in Scandinavia during the 536 disasters. People were willing to part with hugely important items—both in monetary value and social status. Axboe (1999) says it best: "It must have been a very serious matter to sacrifice such things: They were not only valuables, but also expressions of one's social role and status, as well as powerful protective amulets." The community's willingness to part with these items, especially to such a dramatic extent, shows just how dire the circumstances were and how desperately the community needed change to come.

THE ORIGINS OF THE *FIMBULVETR*/FIMBULWINTER MYTH

For decades, historians have been trying to track down where the Fimbulwinter myth came from and if it was based on actual events or not. The term *Fimbulvetr* translates in English as the "great winter." Recently, historians have found connections between the *Fimbulvetr* and the events of 536 AD, believing that the Scandinavian experience of the disaster may have sparked the myth.

In the myth, *Fimbulvetr* is what happens just before the end of the world. Three winters occur with no summer in between them. During this time, the sun won't shine. The myth says that it will finally be swallowed by the wolf who chases it, and the same will happen to the moon. An earthquake will cause enormous destruction, and animals will invade settlements. A sea serpent will cause rising water and tidal waves that also wreak havoc on the people (Sturluson, 1200/2001).

Throughout this time and after, there is chaos and war.

All of this happens as the build up to the *Ragnarök*, which roughly translates in English as the "twilight of the gods." In the *Ragnarök* myth, the gods find out that they are going to be destroyed alongside all life. Even as gods, there is nothing they can do to stop it. After a great battle, it's said that the world will end in fire, smoke, and steam while the Earth falls into the sea.

When all of this subsides, the Earth will come out of the sea and be fertile again. Several of the gods will survive who will restart the world.

Looking back on these myths with what we know about the eruption, it's easy to see the similarities. The eruption of 536 AD would have caused an ongoing winter that lasted for multiple years. This would account for the three winters without summer that are described in the myth.

Similarly, increased humidity and precipitation led to overflowing rivers and higher water levels. Accompanied by more frequent storms as a result of climate change, this could account for the tidal waves and rising water attributed to a sea serpent in the myth.

The absence of the sun and moon in the myth would also make sense in the aftermath of the eruption. We know that the sun was blocked out by the volcanic ash and shone "dimly" like the moon, as described by a Mediterranean historian (Cassiodorus, 1886). Since Scandinavia was even closer to the eruption, they might have seen a more pronounced effect. We can also assume that the myth was slightly exaggerated and that the sun wasn't completely gone—just much dimmer than usual.

Even the earthquakes and animal invasions would make sense in theory with the eruption. Earthquakes are frequently caused by volcanic eruptions, and vice versa (Australian Museum, 2021). Plus, sudden climate changes can disrupt the movements of animals as we saw with the rats that caused the plague. Keep in mind that we have no proof that these things actually happened during the 536 AD event, which means that they were probably just part of the myth. However, they may line up with other historical events, especially since Scandinavia is a region with lots of volcanoes. It's interesting to see that there are possible connections—even with parts of the myth that weren't based on real events in 536 AD.

The end of the world in fire and water could also fit with the eruption, as the volcano spewed lava, and the water levels rose. However, to me, this is a bit of a stretch. As far as we know, the Scandinavians didn't know it was the eruption that started the whole chain of events. It's likely that this part of the myth is based on the eruptions that were common in Scandinavia—not on the 536 one.

The Importance of Fate

One key aspect of these myths that says a lot about Scandinavian perception of the 536 events is fate. There's a lot of discussion around things being "fated" and unavoidable. Fate is so important that even the gods themselves can't prevent it. This was also an important aspect of Scandinavian culture at the time. Like in the *Ragnarök* myth, death was seen as fated and unavoidable. Since nothing could be done about it, being able to die unafraid, or even laughing, was an honorable death (Maraschi, 2019).

In many ways, this part of the myth reflects the circumstances of 536. During the seemingly endless winter and famine, there was nothing that people could really do to stop them. They were stuck with those conditions until they got better and felt that they were at the mercy of the gods to change their circumstances—if they were capable of changing them.

CHAPTER 8

GUPTA AND SASANIAN EMPIRES

Southern Asia and the Middle East also saw dramatic changes as a result of the worst year ever. Due to the plague and shifting political powers, both the Gupta and Sasanian Empires saw great successes and their eventual downfalls.

THE GUPTA EMPIRE

The Gupta Empire was located in northern India from the fourth to sixth centuries. It was known for its vivid culture that was highly influential to the surrounding territories and even overseas. Its extensive trade networks allowed it to have a large sphere of cultural influence. In particular, it was known for its paintings, sculptures, and architectural style. It's often said that India had its golden age under the Gupta Empire.

A BRIEF HISTORY OF THE GUPTA EMPIRE

The Gupta Empire began as a small empire in northeastern India under the rule of Chandra Gupta I and expanded over the generations to cover more territory. The empire grew dramatically under Chandra Gupta's son, Samudra Gupta. He successfully conquered many of the neighboring kingdoms. Contemporaries credited him with defeating at least 20 rulers in the area (Agrawal, 1989). Samudra's successful conquests expanded the Gupta Empire nearly to its peak in terms of territory. At this point, it extended along the eastern shore of the Indian subcontinent and northwest into the region around New Delhi. This gave the Gupta Empire control of the huge amount of wealth that these northern regions had accumulated and increased manpower to improve production and the army.

The empire was expanded once again by the third ruler, Chandra Gupta II. He conquered the neighboring kingdoms along their southwest border and gained control of the western Indian coast. This was an important trading development because it gave them access to seaports in the West. The ability to trade more easily with Western tribes and kingdoms overseas, particularly the Roman Empire, was hugely beneficial and helped the empire to prosper.

Chandra Gupta II also valued education and the arts highly. He encouraged the production of beautiful buildings and artistic works. Together, improving trading and a focus on culture helped the Gupta Empire to thrive and develop its image as a cultural center. Under Chandra Gupta II, the empire achieved its greatest heights and longest period of stability. This is also when the iconic carved stone religious figures emerged, including both Hindu and Buddhist figures.

What would it have been like to live in this Empire, though? Despite the huge successes in the arts and in conquests, life for the average person wasn't amazing in the Gupta Empire. It was pretty equivalent to life in Europe during this time.

That's because, like in Europe, a large portion of the population lived as either slaves or peasants. As expected, the slaves lived in harsh conditions, doing difficult manual labor for their masters. Peasants were free people with much better lives than the slaves. However, they still had to give monetary "gifts" to the emperor and participate in his projects, like constructing new buildings.

As in Rome, the power of the Gupta Empire came from their slave and peasant labor. So, although the empire was thriving, the peasants might not have been.

Fall of the Gupta Empire

Conflict

At this point, the empire began to decline. Under the next ruler, Kumara Gupta, neighboring tribes began to threaten the empire's success. The most notable of these were the Hunas, or Hephthalites, who invaded from the northwest. The Gupta were able to suppress these attacks but doing so used a lot of their resources without netting them any advantages. These attacks continued throughout Skanda Gupta's reign, as well. By the 500s, most of the empire's northwestern region had been taken over by the invading forces (Eraly, 2011).

This had a huge impact on the empire—beyond its toll on the military. The loss of the northwest region dramatically reduced their access to trade, which further reduced their income. The empire couldn't easily access the sea to trade with the Roman Empire, which had been an important trading partner. It also decreased their access to the slaves and peasants who lived along the northwest coast, reducing their manpower and tax income.

These conflicts also hurt the empire's previously thriving culture.

Cultural Declines

As the Huna (or Huns) gradually took control of India through repeated invasions of the Gupta empire, culture fell into decline. This was partially because the empire's economy was no longer thriving, and it struggled to support their military and other basic necessities. They couldn't afford to fund the same amount of art and architectural projects.

Unfortunately, there was also a more pressing reason for their decline. When the Huns invaded, they destroyed monasteries and killed monks in the name of their king, Mihirakula. Mihirakula was notoriously anti-Buddhist and practiced Shaivism, a branch of Hinduism. Both Buddhist and Hindu texts describe him unfavorably, calling him violent and cruel (Pletcher, 2013; Daryaee, 2021).

The destruction of these important religious sites and the death of monks made people afraid to make art that reflected similar values. Since a large amount of the art was religious, particularly the sculptures, people in the empire produced much fewer of these figures.

Over time, both conflict and cultural losses contributed to the decline of the Gupta Empire. The city of Taxila was destroyed, which was an important learning center. This brought a period of cultural regression (Eraly, 2011).

FLOODING

Recent research claims that the final downfall of the Gupta Empire was a huge flood that occurred in the mid-500s (Sharma, 2019). According to the research, no one lived in the area for hundreds of years after the event, showing how devastating the effects of the flood were.

The flooding also had an impact on the decline of Buddhism, as many important Buddhist monasteries were destroyed in this flood (Sharma, 2019). However, a lot of important questions still remain unanswered. We're not sure if the Indian population also experienced famine as a result of the floods, although we can assume that many crops were destroyed by the event. The research also provides several possibilities for the flooding, including an earthquake that caused rivers to change course or extreme rainfall. Either of these could have occurred as a result of the 536 events. However, the conflicts and cultural declines may have been influenced by them, too. We'll discuss this theory more in the "Political Actions and Climate Change" section.

The last ruler of the Gupta Empire was Vishnu Gupta between 540 and 550 AD. The exact end of the Gupta Empire is cloudy, but we know that most of it was eventually taken over by the Aulikara dynasty around 532 AD. The surviving Gupta Empire was very small and weak comparatively until it eventually disappeared. This was probably the result of the floods mentioned earlier.

The Sasanian Empire

The Sasanian Empire was around for four centuries—from 224 to 651 AD. It was the last Iranian empire before the Muslim conquests and became a power strong enough to oppose the Roman Empire. Although it was an Iranian empire, its territory wasn't limited to modern-day Iran. At its peak, it extended out in all directions over parts of the neighboring countries, including the coast of the Arab Peninsula.

The empire was considered one of the peaks of Iranian history with an effective government, educational system, and strong culture. It inspired medieval art throughout the continent and became the foundation of Islamic culture (Daryaee, 2021).

A Brief History of the Sasanian Empire

Before the Sasanian Empire, the Parthian Empire reigned in the same region. It gradually declined as a result of internal issues and ongoing conflict with the Roman Empire.

The Sasanian Empire began with Ardashir I, who gained power as the Parthian Empire was weakened. He was able to gain control of the region by killing the Parthian King Artabanus IV, at the Battle of Hormozdgan in 224 (Procopius, 542 AD/1998).

His first few years as king consisted of continuous rebellions throughout the empire, but he was able to suppress them and continue to expand the territory. During the early years, the empire grew dramatically to the east.

After Ardashir I, his son Shapur I ruled. He also focused on expanding the empire but more so to the west than the east. He led several wars against Rome, most of which were extremely successful and allowed him to expand into the Roman Empire. However, he lost all the territory again after a particularly impactful failure, so all the wars ended up being for nothing.

Eventually, the empire was ruled by Shapur II, and that's when it had its golden age. Empowered by the weakened Roman Empire in the mid-300s, the emperor was able to claim a large amount of the eastern Roman territory for his own. This is also when the empire began to expand down the shore of the Arab Peninsula. It reached one of its largest points under Shapur II's rule.

From Shapur II's death until around 500 AD, the region was relatively peaceful. There were only a few short conflicts with the Roman Empire, and the Sasanian Empire was relatively strong and stable. The two empires developed a treaty of "eternal peace" in 532 (Daryaee, 2021).

Apparently, eternal peace is only eight years long because in 540, ruler Khosrow I broke the treaty and invaded the Roman Empire. The success of this invasion was partially due to the famine and plague that struck Constantinople, forcing the empire to deal with internal issues and weakening their armies. The same plague wouldn't strike the Sasanian Empire until years later.

Along with his military strength, Khosrow I was also known for his focus on knowledge. He supported the Academy of Gondishapur, which collected books from all the surrounding regions and translated them into Persian, Greek, and other languages (Daryaee, 2021). It was an advanced center for learning, offering education and training in medicine, philosophy, theology, and science.

In the early 600s, the Sasanian Empire reached its largest. Ruler Khosrau II ran an incredibly successful military campaign that allowed him to capture Egypt and Turkey from the Romans. This success was short-lived, though.

Fall of the Sasanian Empire

Over time, continued conflicts with the Romans were unfavorable for the Sasanian Empire. They combined with invasions from the Arabs to weaken them.

After the incredible success of the early 600s, the Sasanian Empire was at its largest but also its weakest. It had exhausted its resources expanding so quickly and had raised taxes exorbitantly to fund their military.

These issues combined to make the empire economically and politically unstable. Khosrau was eventually overthrown by his son Kavadh II, who immediately ended the war with Rome and returned all the territories they had gained. After his death, the Sasanian Empire became even more unstable. Multiple kings took power over only a four-year period, and eventually, power fell to the generals. The empire was still in disarray by the time the Muslim conquests occurred.

The Muslim Conquest

The Muslim conquests reached the Sasanian Empire in 633. At this point, the empire was weaker than it had ever been. Some scholars argue that the empire was weaker than any other empire had been in the history of the Persian region (Daryaee, 2021). This made it an ideal time for the Muslims to invade.

Although the empire was weak, it did not give up easily. Battles continued for almost 20 years before most of the region fell into Muslim hands. Even when they did gain control of the territory, frequent and violent uprisings made it unstable. Local people even successfully assassinated some of their Arab governors. It took many years for the area to slowly transition to Islam. Over time, the old Zoroastrian religious texts were burned, and priests were killed. Converting to Islam provided benefits in terms of social status and was probably seen as safer than practicing the old religion (Keys, 2000).

By the late Middle Ages, Islam was the most common religion in Iran.

Political Actions and Climate Change

The Gupta and Sasanian Empire were two powers that, at first glance, might not seem impacted by the events of 536 AD. Their downfalls were mostly due to internal and external conflict—not terrible famine or the plague. So how did the eruption have any impact on these empires?

Environmental Factors

It's possible that these empires were suffering from the widespread climate change that occurred in 536 AD, and we simply don't have a lasting record of it. This may especially be the case for the Gupta Empire.

We have new evidence that proves large areas of the Gupta Empire were flooded just before the time of its collapse. The flood destroyed towns and monasteries, making the area uninhabitable for hundreds of years (Sharma, 2019). Flooding could easily have been caused by the climate changes accompanied by the eruption, as many regions globally saw either droughts or flooding. Sharma (2019) also proposed that an earthquake may have struck that changed the direction of river flow, creating flooding. This could also have been caused by the eruption, but we have no evidence that an earthquake occurred.

At this point, though, the Gupta Empire was already weakened by their conflicts with neighboring powers. The flooding may have been the last straw, but you could argue that the empire's fate was already sealed.

In the case of the Sasanian Empire, there's little evidence of environmental factors. However, we know that the empire was severely affected by the plague, which struck in 627. The plague was able to spread so quickly because the famine weakened people's immune systems throughout Europe and the Mediterranean.

This is the way that many environmental factors influenced the Gupta and Sasanian Empires. We don't have proof that they experienced direct fallout from the eruption like famine or extreme cold, but we do know that they were indirectly affected by these impacts on their neighbors, trading partners, and enemies.

How the Eruption Caused Conflict and Uprooted Empires

The Plague and the Weakening of Major Powers

The plague was a huge factor in the dissolution of empires during this time. It took a particular toll on the Roman and Sasanian Empires, making it difficult for them to manage basic tasks inside their borders. Things like agriculture and production suffered. Huge losses of life meant weaker armies that couldn't achieve their goals. Powerful nations that relied on slave labor from conquered territory suffered because they could no longer sustain their slave population. Poor people were especially vulnerable to the plague because they were more likely to be weakened by famine and have less people to look after them. This created a cycle that hurt these empires more and more over time.

The weakening of the Roman Empire also had an indirect effect on its trading partners including the Gupta Empire. This is because the Romans didn't rely on the Gupta Empire for necessities like grain. Most of their imports from India were luxury items, like silk, iron goods, and spices (Keys, 2000). The Roman Empire was weakened and couldn't afford to buy the same amount of these goods from India. Stagnating trade with the Romans would have been a huge economic disadvantage for the Gupta Empire.

War was often seen as a necessary way to gain back the laborers that had been lost to the plague. Leaders saw the plague as an opportunity to strike while their enemy was weak— not realizing how much they had been weakened themselves. This is what happened to the Sasanian Empire. They expended all their resources to attack the Roman Empire while it was weak, only to lose it all and completely collapse. The same thing nearly happened to Justinian who was dedicated to reconquering Rome almost to his downfall.

Migration

At the same time, the drastic climate change caused by the eruption made people migrate. We saw this in the Americas, Britain, and the Gupta Empire. In these cases, famine, flooding, and other climate issues made it necessary for people to move into other regions.

The destruction of settlements and monasteries devastated an already weakened Gupta Empire, forcing them to recede.

In other cases, tribes and kingdoms saw an opportunity to expand into previously unconquerable land. Areas that had once been well-defended by powerful empires were weak and poorly maintained. This gave them a chance to invade these regions and take the land for themselves. We saw this both with the Gupta and Sasanian Empire.

In the Gupta Empire, invasions from the Huns and other groups might have been either necessitated or allowed by what was going on in the region.

We don't have evidence of famine, but we do know that there was flooding. There may have been other storms and climate events happening in the region that pushed these groups to invade the Gupta Empire. On the other hand, the weakening of the empire due to less trade and failed conquests may have allowed them to invade.

The combination of necessity and opportunity made this an especially busy period for migration, which weakened powerful empires and created new ones in their wake.

Cultural Changes

Widespread disaster took its toll on culture in every corner of the world. Powerful nations couldn't afford to invest in previously vibrant cultures when they could hardly maintain their borders and keep their people from rioting.

The stress of the disasters also led to religious and cultural changes, as people converted in the face of what seemed like the end of the world or were conquered by people with a different religious background. We saw this particularly in the Gupta Empire, where the invading Huns were anti-Buddhist and destroyed all the monasteries and religious artifacts in their wake.

This also happened in the Sasanian Empire when it slowly dissolved and was taken over by Muslim conquests.

Chapter 9

What's to Come, and What Can We Do About It?

Throughout this book, we've hinted that the events of 536 AD reflect modern circumstances in some alarming ways. In many ways, the fallout from the 536 eruption acts as a warning of what we could expect moving forward. Let's go over exactly how and why that's the case, and what we can do to stop it.

How the Climate Impacts Volcanoes, and Vice Versa

In 536 AD, a massive volcanic eruption created a huge chain of events that changed the world. You might think that's one way that 536 and the 2020s are different. We might be facing a climate crisis, global pandemic, supply chain disruptions, and political upheaval, but at least we aren't dealing with an eruption that sends us into a mini ice age… right?

Well, not quite. At least not for now.

Here's the problem: We know that a warmer climate is associated with more volcanic activity, and a cooler climate is associated with less (Sneed, 2017). Researchers have concluded that this is because global warming melts ice in our glaciers. Glacier ice puts pressure on Earth's surface that reduces the flow of magma underground. When that ice melts, the magma that was blocked up is free to go through again, which can cause eruptions (Voosen, 2021).

Similarly, glacier ice can keep unsteady mountains and volcanoes from collapsing. When it melts, long-awaiting landslides can happen. The landslide will relieve pressure just like the melting ice, making the volcano more likely to erupt (Gabbatiss, 2018).

Climate change also makes large volcanic eruptions more impactful because the ash and dust will rise higher into the atmosphere and spread more quickly around the world. This could increase the global cooling effect of these major eruptions by up to 15% (Collins, 2021).

So which volcanoes do we have to worry about right now? Here's a list of volcanoes to watch out for.

Volcanoes Around the World

Katla, Iceland

Katla is an active volcano that is located beneath a glacier in southern Iceland. This one is at a high risk of a large eruption, especially since it sits under a glacier. Melting ice could cause the Katla volcano to erupt, creating a huge disaster much like the one seen in 536 AD (Rafferty, 2015). As an Icelandic volcano that's been around for a long time, it could have even been the cause of the worst year ever. Who knows?

Katla's position underneath a glacier is also alarming for another reason. When a volcano underneath a glacier erupts, it makes the ice evaporate, creating tons of steam and ash (Rafferty, 2015). This can make the eruption even more explosive and increase its impact on the surrounding region.

Yellowstone Caldera, USA

Often called the Yellowstone Supervolcano, this volcano located in Wyoming is known for its massive eruptions throughout history. Scientists previously believed that it wouldn't have another super-eruption for a long time because it has such a huge magma reservoir inside it, and it was nowhere near full.

However, new evidence has proven that these magma reservoirs can fill up suddenly. It doesn't always take centuries for volcanoes to fill up with magma and erupt. Sometimes, it can only be a matter of decades (Aceves, 2017). So, while scientists don't currently expect this volcano to erupt, that could change rapidly as our climate warms.

A Yellowstone Caldera super-eruption would be devastating. Much like the Icelandic eruption in 536, it would release huge amounts of ash that would destroy farmland and kill plants and animals across the whole continent—if not the world (Aceves, 2017). It would cause a global disaster on a level that we haven't seen in a long time.

Mount Fuji, Japan

Mount Fuji is a famous volcano located in Japan, just southwest of Tokyo. It hasn't erupted since the 1700s, but an eruption might be imminent. After the earthquake and following tsunami in 2011, scientists predicted that it would erupt in the next three years (Clark, 2012). It still hasn't happened, meaning that Mount Fuji is overdue for an eruption!

Mount Fuji is also partially collapsing. An earthquake or tsunami in the region could easily trigger the collapse of the volcano, which would be a disaster in itself (Clark, 2012). As the climate warms and storms become more common, this is even more likely.

Mount Shasta, USA

Mount Shasta is an active Californian volcano and the fifth tallest volcano in the United States (NASA, 2018). It's also partially covered in glaciers. Much like Katla, it's a volcano to watch as the melting glaciers could cause an eruption. The area is also known for earthquakes and avalanches, which could trigger (or be triggered by) an eruption.

Fourpeaked Mountain, USA

Fourpeaked Mountain in Alaska is another volcano that is mostly covered in glaciers. It had its last eruption in 2007, a relatively small one mostly characterized by gas release. It's located in an area prone to seismic activity and underneath a glacier—both of which are cause for concern.

Any of these volcanoes, and many more that weren't listed, could erupt as a result of melting glaciers, earthquakes, or storms that cause landslides. As temperatures rise and all of these things become more common, volcanic eruptions are more likely.

Volcanoes: Are They the Solution to Global Warming?

Some people have speculated that volcanoes might be the natural solution to global warming. After all, you just read that global warming can increase the cooling effect of large eruptions, sometimes by as much as 15% (Collins, 2021). As eruptions get bigger and have a greater cooling effect, this could naturally solve global warming! Right?

If only it were that simple. Yes, the volcanoes will have a much greater cooling effect than they previously did. However, they still won't be able to compete with global warming (Voosen, 2021). This gives you some perspective on just how drastic the effects of global warming will be if we continue to let them grow.

Here's a great comparison to help you understand. Just like volcanoes can produce ash and dust that lead to global cooling, they also produce carbon dioxide. Well, humans still produce about *100 times* the carbon dioxide that volcanic eruptions produce (NASA, 2019b). Human carbon dioxide emissions completely dominate volcanic ones. To compete with us, there would have to be a volcanic super-eruption every year, but these super-eruptions usually only happen about once every 100,000 years.

So, while volcanoes are a factor in climate change, scientists predict that they'll be impacted by global warming more than they contribute to (or stop) it.

Supply Chain Disruptions

As a result of the COVID-19 pandemic, we've had some issues with transporting goods. We've seen shipments get stuck in harbors and store shelves stay empty for weeks. What's going on there, and how is it related to what happened in 536 AD?

The supply chain issues we've had with the pandemic are similar to what happened throughout Europe during the plague, especially in highly populated regions like Constantinople. Remember how laborers weren't able to plow the fields because there were simply too many sick people? That's almost exactly what happened with COVID-19.

Many working people quit, were laid off, or became too sick to work. The number of working people has fallen by about 4.7 million since the pandemic began in the U.S. alone (Bhattacharjee et al., 2021). Many of those people don't plan on returning to work—even when the pandemic ends.

At this point, the need for workers exceeds the working population. Job openings have skyrocketed since the pandemic, but there aren't enough people to fill them. At the same time, the people looking for work see a lot of these jobs as a bad deal. The pay is low, and risk has only gone up. People are afraid of getting exposed to COVID-19, and working a job on the front lines requires more time, work, and rules than ever before. Many of these jobs require new policies around cleaning and sanitization that increase workload without extra compensation.

Businesses are struggling to fill vacancies, meaning that some tasks take much longer to complete—if they get done at all. This has been the case in our supply chain while companies struggle to fill vacancies across every step of the process.

The obvious solution would be for these companies to provide better compensation in terms of pay, healthcare, or other benefits. However, most companies aren't willing to pay that price.

As so many companies are suffering from this problem, they're able to get by with fewer workers. Every company is providing less and slower service, so they're still able to keep up with the competition. They cut costs by hiring less employees.

In the end, the people are the ones who suffer. Unstocked shelves and unopened shipments are a common issue that keeps the average person from getting what they need. The items that are available are more expensive, and working people are doing more work for the same wages. Meanwhile, companies continue to cut costs and raise prices in an attempt to keep up with their losses. This is a prime example of history repeating itself as working class people continue to shoulder the biggest burden—even in modern circumstances.

Supply Chain Issues and Environmental Damage

Today, almost everything you need can be bought online and shipped to your door. This was a blessing in the pandemic. Many people opted to order online to avoid exposure to COVID-19, which kept everyone a little bit safer. It was the better option when many of the stores were half empty anyway.

Plus, things like events, concerts, and outings were either canceled or frowned upon. Many people who used to spend money on these experiences were opting to spend it on home decor, clothing, or home hobbies. Objects became one of the main ways that we could spend our extra cash without putting ourselves at risk. This only added to the online ordering frenzy.

For most people, this transition happened automatically. We saw the convenience of two-click orders that showed up at our door, but we didn't see the complicated system that got them there. Combined with the labor shortage, this only made matters worse for the supply chain.

You're probably wondering what all of this has to do with the environment. Actually, it's a lot.

Almost all of the products we buy are made overseas in countries like China and Bangladesh. These products need to be shipped over to North America on cargo ships. On both sides of that process, we need workers. But, at the height of the COVID-19 pandemic, workers were hard to come by. They were often out sick or dealing with mandates and restrictions that led to less people at work and slower work overall. The pandemic greatly reduced the number of workers available to create the goods and get them shipped over to us.

However, the issue was even more pronounced on the U.S. end. Most cargo ships dock in one of two major ports: the Port of Long Beach, or the Port of Los Angeles. These two ports are where 40% of imported goods come into the U.S. (Canon, 2021). At the peak of the pandemic, there was a bottleneck at this point in the process. There weren't enough workers to unload the ships, get the cargo onto the trucks, and then drive those trucks full of product to their destination.

For over a year, there were consistently 40–60 ships idling in the ocean just outside of these ports, waiting for their chance to be unloaded. Sometimes, as many as 70 huge cargo ships were idling just off the coast (Canon, 2021).

These idling ships produced literal tons of pollution. More than 100 tons of smog were released by all these ships, mostly impacting the poor neighborhoods surrounding the docks. This caused health issues including higher rates of asthma, cancer, and even increased death from COVID-19 in these neighborhoods (Canon, 2021). These health effects display the devastating impact that supply chain issues have on people and the environment.

While people on land suffered from worse air pollution, water pollution also hurt marine life. In this way, the pandemic only added to the effect that our supply chains have on the environment.

However, pre-pandemic levels were still extremely high. Before COVID-19, cargo ships like these accounted for about 3% of total greenhouse gas emissions and were projected to go as high as 20% by 2050 (Jones, 2019). This number went up dramatically during the pandemic, but the numbers were already a cause for concern.

Greenhouse gas emissions like the ones from our cargo ships and factories are a big factor in global warming. If nothing is done, we can expect their impact to keep getting worse over time.

Political Disruptions in Times of Strife

From what you've seen in the pandemic, you know that political issues tend to be exacerbated by global crises. People are scared, which makes them more likely to develop an "us versus them" mentality. This sort of mentality keeps them from seeing eye to eye with people who have different perspectives.

It increases tension as the disaster at hand worsens existing issues and exposes new ones. People become more desperate for change, which can lead to protests and riots, then counter-protests and counter-riots.

As we saw in 536 AD, political unrest makes a system more unstable. It was frequently a factor in the fall of powerful empires.

In modern times, a large amount of civil unrest shows that people are dissatisfied with the leadership, regulations, or management of their nation. It can be a sign of civil war, or it can be a normal and healthy part of a nation's life. It depends on if the unrest can be handled in a way that satisfies the population.

We can expect to see more political disruption as we move forward. Climate change notoriously coincides with pandemics and illness, and we've already seen how political tensions are worsened by a pandemic. Plus, storms, natural disasters, and shortages will become more intense and more frequent. These issues will stretch our societies and our governments as they struggle to find adequate solutions alongside economic decline.

I know that all of these changes sound scary, so what can we do to keep things from getting that bad?

What Can You Do?

First off, let's be clear about what the heart of the issue is: climate change. It's already causing drastic changes to our world, and it will only get worse if we don't do anything about it. We'll experience worse storms more often, melting ice caps, flooding, and all the other alarming stuff that comes with global warming.

So what do we do about it?

The problem is that, as people who don't own large corporations or run a powerful country, we can't do that much individually. Most of the biggest contributors to corporations are multimillion-dollar companies that are trading sustainability for higher profits.

For example, here's a fact about climate change: Greenhouse gas emissions by humans contribute more to global warming than any other factor. When you read that, you might feel guilty about driving your car to and from work and think about taking public transit instead.

Doing that is noble, and if everyone can do it, it will make a big difference. But the thing is that canceling your daily drive to work isn't going to stop climate change on any major level. Forcing companies to have sustainable practices *will.*

That doesn't mean that you shouldn't make an individual effort. Instead, it means that you need to put the responsibility for climate change where it belongs: in the hands of the huge corporations that are tossing aside Earth's future in the name of cutting costs.

Your drive to work might burn some fossil fuels, but think about the literal tons of fossil fuels that are burned by cargo ships crossing the ocean every day. What about all the massive trucks burning fuel to bring those goods to the stores? Every one of those items is probably packaged individually in plastic and cardboard inside a bigger box packed with plastic and cardboard. Then there's the factories, and the warehouses, and the fuel to heat and power those factories… Do you see what I mean?

Making changes as an individual is important and necessary, and we'll go over some ways that you can do that. First, though, let's look into how you can hold companies and your government responsible for climate change.

Make Them Listen

Petitioning Your Government

One way that you can push for change is by petitioning your government to implement policies that reduce emissions and improve sustainability. For example, write a letter to your government to urge them to implement greener policies. You can also sign petitions that will put pressure on them. A quick online search will bring up petitions that are relevant to your region.

Boycotting Major Offenders

Another way to push for change is to boycott big-box stores and online marketplaces that are known for unsustainable practices. Encourage others to join the boycott, too! Basically, support businesses that do better, and ditch those that don't.

This is a great tactic because it hits corporations where they'll really feel it: their profits. Companies are more likely to take action when they realize that their unsustainable practices are making a difference in their profits. If you're supporting their sustainable competition instead, they're more likely to do something about their bad practices.

Sometimes, boycotting these companies is easier said than done. These companies tend to offer the lowest prices because they cut costs by underpaying their workers and using cheap, unsustainable energy sources. Keep this in mind and buy sustainably when you can, but don't feel bad if it isn't always possible.

They might seem small, but these are two of the most important things you can do to reduce the effects of climate change. By pushing the biggest offenders to reduce their emissions, we might be able to keep global warming from getting worse.

So, now, you've petitioned your government and are doing your best to buy from sustainable businesses. What else can you do?

Making a Difference on Your Own

Sustainable Travel

As a society, we sure do a lot of driving. Transportation is one of the two biggest ways that the average person generates emissions, and most of that comes from powering our vehicles. Airplane travel also adds up.

One of the best ways to reduce emissions is to make sustainable travel a priority. This can mean taking public transit instead of driving a car or carpooling with other people. Instead of driving short distances, walk or bike there. Trying to minimize plane travel is also helpful for the environment.

Although airplanes do use a lot of energy, they're actually not as bad as you might think. Rather, cars are worse. The average one-person car trip gives off about three-quarters of the emissions that a plane trip would—per person and per mile (BBC News, 2019). Since we tend to take our cars everywhere without a second thought, the fact that driving is nearly as bad as flying is a big deal.

Buses are also surprisingly high offenders. Driving a car with four passengers emits less than half the energy of taking a city bus. This shows the biggest factor in sustainable travel when it comes to vehicles. Driving a car with just one person in it isn't very sustainable, but a full car is much better (BBC News, 2019). That's why carpooling is such a great option.

Meanwhile, trains and long-distance buses (like coach buses) are the most sustainable choices, carrying more people further with less emissions.

Housing

Surprisingly, housing is also a big factor. It's one that most people don't think about, but everyone has to live somewhere. Living inside involves basic necessities like running water, heating, air conditioning, and electricity. All of these things use energy, and most of it isn't obtained in a sustainable way. Energy companies are some of the biggest offenders when it comes to unsustainable practices (Crabtree, 2021a).

There are a lot of simple ways to reduce your carbon footprint at home. Let's go through the list (Crabtree, 2021a):

1. Insulate your home. This will keep your house at a stabler temperature, so you won't have to spend so much energy (and money) heating or cooling it.
2. Turn off what you aren't using. This applies to lights, electronics, and anything else that uses energy.
3. Replace old electronics with energy efficient ones. The next time an appliance dies, look for an energy-efficient option, so you can do the same task using less energy.
4. Be thoughtful about plastic use. Try to avoid introducing products into your home that come wrapped in or made of plastic whenever possible. Plastic is made from fossil fuels and doesn't break down. It's especially important to avoid single-use plastic, like cups, utensils, wrappers, and bags.

Food

Not all food is created equal. Some foods require significantly more energy to produce and create more emissions. Beef and lamb are the worst, but all kinds of meat are incredibly bad for the environment (Crabtree, 2021b).

That's partially because livestock produce methane. However, you also need to have outdoor space for animals like cows to live on. This has been a big cause of deforestation as we try to clear more land for these animals to eat and live on. We also need more land to grow the crops for those animals to eat, which means more deforestation (Crabtree, 2021b).

Meat also requires a lot of water—both for processing and to feed to the livestock. Plus, as these animals eat the grass, they leave the land bare. It eventually will become unfertile, meaning nothing can be grown there later.

The worst meats in terms of environmental impact are lamb, beef, pork, and chicken.

Although not a meat, cheese also has a surprisingly high impact just after the red meats.

In general, dairy products and eggs are also good to avoid.

Eating a vegan diet is the best thing you can do for the environment, but that doesn't mean you need to go vegan right away. If a dramatic diet change doesn't sound right for you, you can always take small steps by reducing your consumption of red meat over time. Big changes like this are best made slowly—not all at once!

Consumerism

The last major factor in individual carbon emissions is consumerism. As much as we don't want to admit it, we all like to buy stuff. Paying attention to what you're spending money on is good both for your bank account and the environment. A lot of the things we buy just end up in the trash. If you buy too much food, clothing, or other items that you don't really need, you're not the only one. The best thing you can do is just pay attention and try to cut back when you can. It won't be perfect, but it doesn't have to be. Every little bit helps.

Making an effort to buy locally is also important. By supporting small stores in your area and buying groceries from local farmers, you help them stay afloat amid all the big-box stores. Plus, you reduce the need for goods brought in from overseas, preventing all the pollution that comes with that. Avoiding international items is key, but even goods from other counties or states aren't as sustainable. Ideally, you'll get everything you need from local sellers, dramatically reducing the emissions required to transport goods globally. Communities will become more self-sustaining, and we'll be less impacted by supply chain issues.

Supporting your local community will help businesses and their owners to thrive, which helps your local economy and provides other benefits. As we saw in North America during the events of 536, creating a strong local community provides the whole area with a support system. You'll have neighbors and friends to rely on if things go awry. It isn't just the more sustainable option; it's also better for you and your neighborhood.

Working Together Is Key

If one person goes vegan and buys locally, that's an amazing step… but it won't be enough on its own. To create dramatic change, we need everyone to do their part. That doesn't mean that everyone needs to quit red meat today and move into a tiny house. Some sustainable options might not be sustainable for *you.* Instead, just do something small and doable. For example, if everyone did "Meatless Monday," we could hypothetically cut the world's consumption of meat by one seventh. That's huge!

The point is that if everyone does what they can, it will make a difference. Start with a small step that's manageable for you, and move up from there. If everyone does their part, we can create real, lasting change.

Conclusion

536 AD Proves That We Need to Change

The parallels between 536 AD and 2020 are alarming to say the least. We saw how a massive volcanic eruption sent the whole world into disarray. It exposed just how similar our circumstances are today and how close we could be to the brink of disaster.

In 536, a massive Icelandic eruption caused global cooling, sending the planet into a mini ice age in early spring. The growing season ceased to exist, and summer frosts ruined crops everywhere from South America to Europe and China. This caused a famine that killed thousands. Those who survived were seriously weakened, making it easy for the plague to spread.

At the same time, cold and famine necessitated migration. Tribes and kingdoms were forced to move in search of food and milder conditions. Other societies saw an opportunity to invade while their neighbors were weak and sick.

The external threats combined with the plague and food and labor issues to create civil unrest. In what seemed like the end of the world, many people sought comfort in religion and culture. However, these aspects of life were changing. Some religions fell out of favor while others flourished. Conflict and influence from neighboring regions changed popular culture in surprising ways.

Recovery was a long and grueling process. Many historians consider 536 the beginning of the hardest time to be alive. It created a chain of events that changed the course of humanity, and it took hundreds of years for the world to fully recover. What happened in 536 AD still echoes in our cultures.

Diving deep into history, it was impossible to miss the connection with present-day events. Labor shortages due to the plague mirror modern supply chain issues. We're seeing political upheaval in the face of uncertain circumstances and a pandemic, too.

Perhaps the most alarming realization of all is the devastating toll that natural disasters can take on humanity. It only took one massive eruption to coat the sky in ash and send the entire world into a downward spiral lasting hundreds of years.

Our world is warming up fast. At the rate we're going, we'll be seeing the most devastating impacts of global warming in only a few years (NASA, 2019a). We already have more frequent and intense storms, and we can only expect it to get worse as the climate warms, and glaciers melt.

We're at a crucial point in history. We still have time to change things for the better and prevent further global warming, but it will take all of us. We need to work together to hold powerful people accountable and force them to see the facts. Remind the big corporations that if there's no Earth, there's no money, either.

Individually, we can all do our part to reduce our emissions and prevent the worst of global warming. Being thoughtful about our energy use, especially fossil fuels, is an important step. We can also reduce our impact by eating more sustainably and trying to buy less.

If we all chip in, we can protect ourselves from a disastrous outcome like what happened in 536 AD.

Don't let 2020 be our 536. What happened in 2020 was a wake-up call, begging us to see that our way of life is precarious and unsustainable. We can't afford to ignore that call anymore. We no longer have the time to make thoughtless decisions or let the situation continue to spiral. The time to act is now, and the only way through is together. Let's all do our part to make the future incredible.

AUTHOR BIOGRAPHY

I'm a history buff and world traveler who is passionate about the lived experiences behind historical dates and figures. I compare the past and present to teach readers about how much we have in common with historical people. I'm endlessly intrigued by the way history repeats itself, and I believe that historical knowledge is an important way to keep humanity from making the same mistakes. I have a lifelong fascination with sociology and anthropology which has given me a deep appreciation for every stage of human progression and the struggles that go along with them. My ultimate hope is to remind my readers that history isn't over: It's how our ancestors lived which, in turn, changes who we are and how we live. We are not as far removed from it as we may feel when we sit in a history class! This realization inspired me to pursue historical writing, so I can share my expertise and worldly experiences with others. I have been to 96 countries so far, and I'm excited to continue learning and exploring. I have a deep love for humanity across all cultures and time periods, and that love motivates me to try and make the world a better place. When I'm not writing or seeing the world, you can find me cuddling with my rescue cats, enjoying quality family time, going for walks, or wine tasting. I also love cooking, painting modern art, and music. I live in Hawaii and have two adult children.

REFERENCES

Aceves, A. (2017). *Yellowstone supervolcano may erupt sooner than anticipated.* NOVA. https://www.pbs.org/wgbh/nova/article/yellowstone-supervolcano-may-erupt-sooner-than-anticipated/

Agrawal, A. (1989). Rise and fall of the Imperial Guptas. In *Google Books.* Motilal Banarsidass Publications.
https://books.google.ca/books?id=hRjC5IaJ2zcC&pg=PA315&redir_esc=y#v=one page&q&f=false

Akram, A. I. (2009). *The Muslim conquest of Persia.* Maktabah Booksellers and Publishers.

https://www.islamawareness.net/MiddleEast/Iran/iran_article0002.pdf

Anderson, B. S., & Zinsser, J. P. (1989). *A history of their own: Women in Europe from prehistory to the present, vol. 1.* Harper & Row.

Andrews, E. (2014, January 14). *8 reasons why Rome fell.* History. https://www.history.com/news/8-reasons-why-rome-fell

Australian Museum. (2021, September 29). *How are volcanoes and earthquakes interrelated?*https://australian.museum/learn/minerals/shaping-earth/how-are-volcanoes-and-earthquakes-
interrelated/#:~:text=Similarly%2C%20volcanoes%20can%20trigger%20earthquakes

Axboe, M. (1999). The year 536 and the Scandinavian gold hoards. *Medieval Archaeology, 43,* 186–188.

Bauer, P. (2018). Mount Tambora: Location, eruptions, & facts. In *Encyclopedia Britannica.* https://www.britannica.com/place/Mount-Tambora

BBC News. (2019, August 23). Climate change: Should you fly, drive or take the train? *BBC News.* https://www.bbc.com/news/science-environment-49349566

Bhattacharjee, D., Bustamante, F., Curley, A., & Perez, F. (2021, December 10). *Reimagining supply-chain jobs to attract and retain workers.* McKinsey & Company. https://www.mckinsey.com/business-functions/operations/our-insights/navigating-the-labor-mismatch-in-us-logistics-and-supply-chains

Büntgen, U., Myglan, V. S., Ljungqvist, F. C., McCormick, M., Di Cosmo, N., Sigl, M., Jungclaus, J., Wagner, S., Krusic, P. J., Esper, J., Kaplan, J. O., de Vaan, M. A. C., Luterbacher, J., Wacker, L., Tegel, W., & Kirdyanov, A. V. (2016). Cooling and societal change during the Late Antique Little Ice Age from 536 to around 660 AD. *Nature Geoscience, 9*(3), 231–236. https://doi.org/10.1038/ngeo2652

Bury, J. B. (1958). *History of the later Roman Empire from the death of Theodosius I. to the death of Justinian.* London: Courier Corporation.

Canon, G. (2021, October 15). *Ships backed up outside US ports pumping out pollutants as they idle.* The Guardian. https://www.theguardian.com/business/2021/oct/15/us-california-ports-ships-supply-chain-pollution

Cartwright, M. (2015). *Muisca civilization.* World History Encyclopedia. https://www.worldhistory.org/Muisca_Civilization/

Cartwright, M. (2018). *Byzantine Emperor.* World History Encyclopedia. https://www.worldhistory.org/Byzantine_Emperor/#:~:text=Aided%20by%20ministers%2C%20high%2Dranking

Cassiodorus, S. (1886). The letters of Cassiodorus. In *Google Books* (pp. 518–520). H.

Frowde.

https://books.google.ca/books?id=aymsvxyyOhoC&pg=PA518&redir_esc=y#v=on

epage&q&f=false

Centers for Disease Control and Prevention [CDC]. (2021, June 23). *Going to a public*

disaster shelter during the COVID-19 pandemic. https://www.cdc.gov/disasters/hurricanes/covid-19/public-disaster-shelter-during-covid.html

Clark, L. (2012). *Pressure in Mount Fuji is now higher than last eruption, warn experts.* Wired. https://www.wired.co.uk/article/mount-fuji

Collins, S. (2021, August 12). *Climate change will transform cooling effects of volcanic eruptions, study suggests*. University of Cambridge. https://www.cam.ac.uk/stories/volcanoesandclimate#:~:text=They%20found%20that%20for%20large

Crabtree, M. (2021a, January 12). *How to reduce your carbon footprint – 20 top tips.* FutureLearn. https://www.futurelearn.com/info/blog/how-to-reduce-your-carbon-footprint-tips

Crabtree, M. (2021b, May 11). *Is eating meat bad for the environment? 6 global impacts.*

FutureLearn. https://www.futurelearn.com/info/blog/eating-meat-bad-for-environment

Daryaee, T. (2021). King of the seven climes: A history of the ancient Iranian world (3000 BCE - 651 CE). Jordan Center for Persian Studies.

Diamond, J. (2019). *The story of smallpox and other deadly Eurasian germs.* Public

Broadcasting Station. https://www.pbs.org/gunsgermssteel/variables/smallpox.html

Eraly, A. (2011). The first spring: The golden age of India. Penguin Books India. https://books.google.ca/books?id=te1sqTzTxD8C&pg=PA48&redir_esc=y#v=onepage&q&f=false

Gabbatiss, J. (2018). *Climate change could trigger volcanic eruptions across the world,*

warnscientists.Independent.http://www.independent.co.uk/climate-

change/news/volcano-eruption-climate-change-mountain-landslide-glacier-global-

warming-a8299821.html

Garland, L. (2002). Byzantine empresses: Women and power in Byzantium AD 527-1204.

In*GoogleBooks*.Routledge.

https://books.google.ca/books?id=6gWGAgAAQBAJ&printsec=frontcover&source=gbs_atb#v=onepage&q&f=false

Gibbons, A. (2018). Eruption made 536 "the worst year to be alive." *Science, 362*(6416), 733–734. https://doi.org/10.1126/science.362.6416.733

Gräslund, B., & Price, N. (2012). Twilight of the gods? The "dust veil event" of AD 536 in critical perspective. *Antiquity, 86*(332), 428–443. https://doi.org/10.1017/s0003598x00062852

Greshko, M. (2019, August 23). *Colossal volcano behind "mystery" global cooling finally found.* Science. https://www.nationalgeographic.com/science/article/colossal-volcano-behind-mystery-global-cooling-found

Halsall, G. (2014). Two worlds become one: A "counter-intuitive" view of the Roman Empire and "Germanic" migration. *German History, 32*(4), 515–532. https://doi.org/10.1093/gerhis/ghu107

Helama, S., Arppe, L., Uusitalo, J., Holopainen, J., Mäkelä, H. M., Mäkinen, H., Mielikäinen, K., Nöjd, P., Sutinen, R., Taavitsainen, J.-P., Timonen, M., & Oinonen, M. (2018). Volcanic dust veils from sixth century tree-ring isotopes linked to reduced irradiance, primary production and human health. *Scientific Reports, 8*(1). https://doi.org/10.1038/s41598-018-19760-w

Higham, N. J. (2021). King Arthur: The making of the legend. Yale University Press. History. (2018, August 29). *Was King Arthur a real person?*

https://www.history.com/news/was-king-arthur-a-real-person

Horgan, J. (2014, December 26). *Justinian's Plague (541-542 CE).* World History Encyclopedia. https://www.worldhistory.org/article/782/justinians-plague-541-542-ce/

Hughes, T. (2021, September 14). *Why was 900 years of European history called "the Dark Ages"?* History Hit. https://www.historyhit.com/why-were-the-early-middle-ages-called-the-dark-ages/#:~:text=It%20has%20been%20called%20the

Jacobsen, T. C. (2012). *The Gothic War: Justinian's campaign to reclaim Italy.*

Westholme Publishing.

Jarus, O. (2013, December 21). *History of the Byzantine Empire (Byzantium).* Live Science. https://www.livescience.com/42158-history-of-the-byzantine-empire.html Johnson, C. Y. (2012). *Native Americans migrated to the New World in three waves,*

Harvard-led DNA analysis shows. Boston. https://www.boston.com/uncategorized/noprimarytagmatch/2012/07/11/native-

americans-migrated-to-the-new-world-in-three-waves-harvard-led-dna-analysis-shows/

Jones, K. (2019, May 7). *The environmental impact of maritime freight.* Dynamo. https://www.dynamo.vc/blog-posts/the-environmental-impact-of-maritime-freight

Kanisetti, A. (2018, April 11). *Volcanic activity and the fall of the Gupta Empire.* Medium. https://nationalinterest.in/volcanic-activity-and-the-fall-of-the-gupta-empire-8c9b463d1663

Keum Young, A., Damsteegt, G., de Kock, E., Sook Young, K., Jhung Haeng, K., Myun Ju, L., Miller, N., Dae Geuk, N., O'Reggio, T., Shea, W. H., Treiyer, A. R., & van Wyk, K. (2017). 538 A.D. and the transition from Pagan Roman Empire to Holy Roman Empire: Justinian's metamorphosis from chief of staffs to theologian. *International Journal of Humanities and Social Science, 7*(1). https://www.ijhssnet.com/journals/Vol_7_No_1_January_2017/7.pdf

Keys, D. (2000). Catastrophe: An investigation into the origins of the modern world. In *The Open Library* (1st American ed.). Ballantine Pub. https://openlibrary.org/books/OL32597M/Catastrophe

Kutterolf, S., Freundt, A., & Peréz, W. (2008). Pacific offshore record of plinian arc volcanism in Central America: 2. Tephra volumes and erupted masses. *Geochemistry, Geophysics, Geosystems, 9*(2), n/a-n/a. https://doi.org/10.1029/2007gc001791

Lapidus, I. M. (2014). *A history of Islamic societies.* Cambridge University Press. Larsen, L. B., Vinther, B. M., Briffa, K. R., Melvin, T. M., Clausen, H. B., Jones, P. D., Siggaard-Andersen, M. L., Hammer, C. U., Eronen, M., Grudd, H., Gunnarson, B. E., Hantemirov, R. M., Naurzbaev, M. M., & Nicolussi, K. (2008). New ice core evidence for a volcanic cause of the A.D. 536 dust veil. *Geophysical Research Letters, 35*(4). https://doi.org/10.1029/2007gl032450

Little, B. (2018, December 3). *The worst time in history to be alive, according to science.* History. https://www.history.com/news/536-volcanic-eruption-fog-eclipse-worst-year

Little, L. K. (2007). *Plague and the end of antiquity: The pandemic of 541–750.* Cambridge University Press. http://www.academia.dk/MedHist/Sygdomme/Pest/PDF/Plague_and_the_End_of_Antiquity.pdf

Loveluck, C. P., McCormick, M., Spaulding, N. E., Clifford, H., Handley, M. J., Hartman, L., Hoffmann, H., Korotkikh, E. V., Kurbatov, A. V., More, A. F., Sneed, S. B., & Mayewski, P. A. (2018). Alpine ice-core evidence for the transformation of the European monetary system, AD 640–670. *Antiquity, 92*(366), 1571–1585. https://doi.org/10.15184/aqy.2018.110

Manzanilla, L. R. (2015). Cooperation and tensions in multiethnic corporate societies using Teotihuacan, Central Mexico, as a case study. *Proceedings of the National Academy of Sciences, 112*(30), 9210–9215. https://doi.org/10.1073/pnas.1419881112

Maraschi, A. (2019). Learning from the past to understand the present: 536 AD and its consequences for man and the landscape from a catastrophist perspective. *Cerae: An Australasian Journal of Medieval and Early Modern Studies, 6.* https://ceraejournal.com/wp-content/uploads/2021/06/Maraschi.-Copyedited.-2.pdf

Mark, J. J. (2020). *Procopius on the Plague of Justinian: Text & commentary.* World

History Encyclopedia. https://www.worldhistory.org/article/1536/procopius-on-the-plague-of-justinian-text--comment/

Metropolitan Museum of Art. (2021). *North America, 500-1000 A.D.* https://www.metmuseum.org/toah/ht/06/na.html

Metropolitan Museum of Art. (2022). *Plate with youths and winged horses ca. 5th–6th*

century A.D. https://www.metmuseum.org/art/collection/search/325650

Miller, G. H., Geirsdóttir, Á., Zhong, Y., Larsen, D. J., Otto-Bliesner, B. L., Holland, M. M., Bailey, D. A., Refsnider, K. A., Lehman, S. J., Southon, J. R., Anderson, C., Björnsson, H., & Thordarson, T. (2012). Abrupt onset of the Little Ice Age triggered by volcanism and sustained by sea-ice/ocean feedbacks. *Geophysical Research*

Letters, 39(2), n/a-n/a. https://doi.org/10.1029/2011gl050168

Miller, M. (2016). *Social inequality and the body: Diet, activity, and health differences in a prehistoric Muisca population (Sabana de Bogotá, Colombia, AD 1000-1400).* https://digitalassets.lib.berkeley.edu/etd/ucb/text/Miller_berkeley_0028E_16639.pdf

Monmouth, G. (1999). *History of the Kings of Britain* (J. A. Giles, Ed.; A. Thompson,

Trans.). Cambridge Medieval Latin Series. https://www.yorku.ca/inpar/geoffrey_thompson.pdf (Original work published 1136)

Mordechai, L., & Eisenberg, M. (2019). Rejecting catastrophe: The case of the Justinianic Plague. *Past & Present, 244*(1), 3–50. https://doi.org/10.1093/pastj/gtz009

Mordechai, L., Eisenberg, M., Newfield, T. P., Izdebski, A., Kay, J. E., & Poinar, H. (2019). The Justinianic Plague: An inconsequential pandemic? *Proceedings of the National*

Academy of Sciences, 116(51), 25546–25554. https://doi.org/10.1073/pnas.1903797116

NASA. (2018, May 20). *Mount Shasta, California.* NASA's Earth Observatory. https://earthobservatory.nasa.gov/images/92174/mount-shasta-california#:~:text=1%2C%202018JPEG-

NASA. (2019a). *Is it too late to prevent climate change?* NASA: Vital Signs of the Planet. https://climate.nasa.gov/faq/16/is-it-too-late-to-prevent-climate-change/

NASA. (2019b). *What do volcanoes have to do with climate change?* NASA: Global Climate Change. https://climate.nasa.gov/faq/42/what-do-volcanoes-have-to-do-with-climate-change/

Newfield, T. (2016, May 1). *The global cooling event of the sixth century. Mystery no longer?* Historical Climatology. https://www.historicalclimatology.com/features/something-cooled-the-world-in-the-sixth-century-what-was-it

Norris, S. (2008, May 12). *Inca skull surgeons were "highly skilled," study finds.* National Geographic. https://www.nationalgeographic.com/science/article/news-trepanation-inca-medicine-archaeology

Paowary, K. (2018). *The collapse of Marib Dam and the fall of an empire.* Amusing Planet. https://www.amusingplanet.com/2018/11/the-collapse-of-marib-dam-and-fall-of.html

Peregrine, P. N. (2020). Climate and social change at the start of the Late Antique Little Ice Age. *The Holocene, 30*(11), 1643–1648. https://doi.org/10.1177/0959683620941079

Perkins, S. (2016, February 8). *Volcano-induced "Little Ice Age" may have contributed to famines,warsin6thand7thcenturies.*Science.

https://www.science.org/content/article/volcano-induced-little-ice-age-may-have-contributed-famines-wars-6th-and-7th-centuries

Pletcher,K.(2013).*Mihirakula.*EncylopediaBrittanica.

https://www.britannica.com/biography/Mihirakula

Procopius (1916). Procopius: History of the wars, books III and IV (Vandalic War). In

https://books.google.com.au/books?redir_esc=y&id=szQjAQAAMAAJ&pg=PA329&sig=ACfU3U3viCnyjDWpIekZMiXINambZi7GiA&focus=searchwithinvolume&q=portent#v=snippet&q=portent&f=false

Procopius. (1998). *History of the wars, II.* Fordham University Center for Medieval Studies.https://sourcebooks.fordham.edu/source/542procopius-plague.asp (Original work published 542 AD)

Rafferty,J.P.(2015).Katlavolcano,Iceland.In*EncyclopediaBritannica.*

https://www.britannica.com/place/Katla

Ray,M.(2014).Ostrogoth.In*EncyclopediaBritannica.*

https://www.britannica.com/topic/Ostrogoth

Ray, M. (2019). Code of Justinian | Definition & Creation. In *Encyclopedia Britannica.* https://www.britannica.com/topic/Code-of-Justinian

Rigby, E., Symonds, M., & Ward-Thompson, D. (2004). A comet impact in AD 536? *Astronomy and Geophysics, 45*(1), 1.23–1.26. https://doi.org/10.1046/j.1468-4004.2003.45123.x

Robock, A., & Toon, O. B. (2016). Self-assured destruction: The climate impacts of nuclear

war. *Bulletin of the Atomic Scientists, 68*(5), 66–74. https://doi.org/10.1177/0096340212459127

Roosevelt, A. C., Lima da Costa, M., Lopes Machado, C., Michab, M., Mercier, N., Valladas, H., Feathers, J., Barnett, W., Imazio da Silveira, M., Henderson, A., Sliva, J., Chernoff, B., Reese, D. S., Holman, J. A., Toth, N., & Schick, K. (1996). Paleoindian cave dwellers in the Amazon: The peopling of the Americas. *Science, 272*(5260), 373– 384. https://doi.org/10.1126/science.272.5260.373

Rosen, W. (2007). *Justinian's flea: Plague, empire and the birth of Europe.* Jonathan Cape.

Sanders, W. T., & Webster, D. (1988). The Mesoamerican urban tradition. *American Anthropologist, 90*(3), 521–546. https://doi.org/10.1525/aa.1988.90.3.02a00010

Sarchet, P. (2016). *125-year mini ice age linked to the plague and fall of empires.* New Scientist. https://www.newscientist.com/article/2076713-125-year-mini-ice-age-linked-to-the-plague-and-fall-of-empires/

Sarris, P. (2021). New approaches to the "Plague of Justinian." *Past & Present, 254*(1). https://doi.org/10.1093/pastj/gtab024

Sessa, K. (2020). *The Justinianic Plague.* Origins. https://origins.osu.edu/connecting-history/covid-justinianic-plague-lessons?language_content_entity=en

Sharma, S. (2019). *Deluge drowned mighty Guptas: Study.* Www.telegraphindia.com. https://www.telegraphindia.com/india/deluge-drowned-mighty-guptas-study/cid/1685500

Sinensky, R. J., Schachner, G., Wilshusen, R. H., & Damiata, B. N. (2021). Volcanic climate forcing, extreme cold and the Neolithic Transition in the northern US Southwest. *Antiquity*, 1–19. https://doi.org/10.15184/aqy.2021.19

Smith, V. C., Costa, A., Aguirre-Díaz, G., Pedrazzi, D., Scifo, A., Plunkett, G., Poret, M., Tournigand, P.-Y., Miles, D., Dee, M. W., McConnell, J. R., Sunyé-Puchol, I., Harris, P. D., Sigl, M., Pilcher, J. R., Chellman, N., & Gutiérrez, E. (2020). The magnitude and impact of the 431 CE Tierra Blanca Joven eruption of Ilopango, El Salvador. *Proceedings of the National Academy of Sciences, 117*(42), 26061–26068. https://doi.org/10.1073/pnas.2003008117

Sneed, A. (2017, December 21). *Get ready for more volcanic eruptions as the planet warms.* Scientific American. https://www.scientificamerican.com/article/get-ready-for-more-volcanic-eruptions-as-the-planet-warms/

Stothers, R. B. (1984). Mystery cloud of AD 536. *Nature, 307*(5949), 344–345. https://doi.org/10.1038/307344a0

Sturluson, S. (2001). *The Prose Edda.* Blackmask Publishing. https://is.cuni.cz/studium/predmety/index.php?do=download&did=62028&kod=ARL100252 (Original work published 1200)

Teall, J. L., & Nicol, D. M. (2018). Byzantine Empire: History, geography, maps, & facts. In *Encyclopedia Britannica.* https://www.britannica.com/place/Byzantine-Empire The Editors of Encyclopedia Britannica. (2018). Mount Tambora: Location, eruptions, & facts. In *Encyclopedia Britannica.* https://www.britannica.com/place/Mount-

Tambora

Tramontana, F. (2013). The poll tax and the decline of the Christian presence in the Palestinian countryside in the 17th century. *Journal of the Economic and Social*

History of the Orient, 56(4-5), 631–652. https://doi.org/10.1163/15685209-12341337

Tunturi, J. (2011). Darkness as a metaphor in the historiography of the Enlightenment. *Approaching Religion, 1*(2), 20–25. https://doi.org/10.30664/ar.67479

United Nations. (2022). *TONGA: Volcanic eruption situation report no. 1* (pp. 1–6).

OCHA Office of the Pacific Islands.

Voosen, P. (2021, August 19). *Massive volcanoes could cool Earth more in a warming world.* Science. https://www.science.org/content/article/massive-volcanoes-could-cool-earth-more-warming-world

Williams, J. M. (1975). Origins of the English language, a social and linguistic history. In

Internet Archive. New York: Free Press. https://archive.org/details/originsofenglish0000will

Wolfram, H. (1990). *History of the Goths.* University of California Press.

Worral, S. (2017, March 5). *Past disasters reveal terrifying future of climate change.*

Science. https://www.nationalgeographic.com/science/article/climate-change-global-warming-history-health

Yalman, S. (2001). *The birth of Islam.* The Metropolitan Museum of Art. https://www.metmuseum.org/toah/hd/isla/hd_isla.htm

Zielinski, S. (2015, February 23). *Plague Pandemic May Have Been Driven by Climate, Not Rats.* Smithsonian. https://www.smithsonianmag.com/science-nature/plague-pandemic-may-have-been-driven-climate-not-rats-180954378/

Made in United States
Troutdale, OR
01/25/2025

28331965R00087